Lecture Notes in Biomathematics

Managing Editor: S. Levin

16

G. Sampath
S. K. Srinivasan

Stochastic Models for Spike Trains of Single Neurons

Springer-Verlag
Berlin · Heidelberg · New York 1977

Authors

G. Sampath
S. K. Srinivasan
Dept. of Mathematics
Indian Institute of Technology
Madras 600036/India

Library of Congress Cataloging in Publication Data

Srinivasan, S Kidambi.
Stochastic models for spike trains of single neurons.

(Lecture notes in biomathematics ; v. 16)
Bibliography: p.
Includes index.
1. Excitation (Electrophysiology)--Mathematical models. 2. Neurons--Mathematical models. 3. Action potentials (Electrophysiology)--Mathematical models. 4. Stochastic analysis. I. Sampath, Gopalan, 1948- joint author. II. Title. III. Series.
QP363.S64 591.1'88 77-2728

AMS Subject Classifications (1970): 60-02, 60G35, 60J70, 60K05, 60K10, 60K20, 60K35, 92-02, 92A05

ISBN 3-540-08257-3 Springer-Verlag Berlin · Heidelberg · New York
ISBN 0-387-08257-3 Springer-Verlag New York · Heidelberg · Berlin

Printing and binding: Beltz Offsetdruck, Hemsbach/Bergstr.
2145/3140-543210

PREFACE

These notes are aimed at collecting in one volume the vast and scattered literature on stochastic models of spontaneous activity in single neurons. An attempt has been made to make the treatment self-contained by providing an introduction to neurophysiology as well as the mathematical background for each kind of model. The mathematical aspects of a model are stated as a series of lemmas and theorems; this gives the relative importance of the different results and also facilitates easy reference. However, the proofs are often sketchy and sometimes omitted. This has been done to make the notes compact and easier to read. While the coverage is fairly wide, not all studies in the literature are described herein, mainly because the differences are slight. This, however, should not be construed as an understatement of the relative importance of some of them; they have been included in a list of additional references. This list also includes studies in fields like operations research, inventory control and reliability theory. These have possible applications in neuron modelling though no direct reference is made to them in the notes.

Many of the figures have been adapted from various publications and permission from the following publishers is gratefully acknowledged: M/s Springer-Verlag, Academic Press, Macmillan Journals Ltd., Applied Probability Ltd., Wiley-Interscience, McGraw-Hill, Cambridge University Press, Society for Mathematical Biology, American Elsevier and American Association for the Advancement of Science.

It is a pleasure to thank Mr.A.V.Chandrasekaran for the care he has taken in typing these notes.

Madras S.K.Srinivasan

31.1.1977 G.Sampath

CONTENTS

INTRODUCTION

In recent years, there have been many significant advances in neurobiology. Experimental investigations have revealed several interesting facts about the nervous system. Simultaneously there have been attempts to develop quantitative theories of behaviour. Basically, the nervous system is the communications system of an organism in which the signals are largely electrochemical, and consists of a network of cells interconnected by fibers. It is natural therefore to study its electrical properties. There are three levels of approach. Thus, one could study

1) propagation of a signal along a fiber,
2) the behaviour of a single cell,
3) clusters of such cells.

Signal transmission along fibers is well understood and comprehensive accounts are available. The electrical properties of single cells are also known in great detail, but, since cells are interconnected, theoretical study of single cells as well as networks of such cells is exceedingly difficult.

It is nevertheless useful to study the behaviour of a single cell by making simplifying assumptions, the principal one being that the cell is isolated. The present work reviews mathematical models which have been proposed in the literature to explain one aspect of 2), namely, spike trains generated by spontaneous activity in single neurons.

The output of a neuron consists of a sequence of voltage impulses. Such a time sequence may be assumed to be a stream of point events, an assumption that is justified in many types of neurons. A neuron may be active even when it is not stimulated. In such a phase of spontaneous activity, the sequence possesses the characteristics of a stochastic point process. This has motivated the use of the theory of stochastic processes in modelling the spontaneous activity of single neurons. Several models based on different assumptions have been described in the literature.

These models are reviewed in the present work. An outline of these notes is given below.
Chapter 1 presents some relevant background from neurophysiology, the treatment being highly simplified. The neuron and its parts are described, the functional mechanisms being emphasised. Structural differences in the synaptic region are also discussed to facilitate relating mathematical models to specific types of neurons.
In Chapter 2, the assumption that a neural impulse sequence is a point process is justified. When the neuron is spontaneously active, the output is a stochastic point process. Examples of spontaneous activity recorded by several investigators are given.
In Chapter 3, the stochastic properties of impulse trains are discussed in the light of the statistical distribution of the interval between two successive impulses, stationarity of the sequence, etc. The properties of the mathematical neuron are then discussed.
In Chapters 4 to 9, the different types of models are introduced and discussed in detail. These are

1) superposition models
2) deletion models
3) counter models
4) diffusion models
5) discrete state random walk models
6) continuous state random walk models.

At every stage the advantages and disadvantages of the models as well as their applicability to specific types of neurons are discussed.
In the concluding chapter, properties of real neurons and the corresponding properties in the model are critically discussed.

A monograph on the stochastic activity of neurons by Holden (1976) has been recently published in this Lecture Notes series while the present work was in preparation. Holden discusses many aspects of spontaneous activity of both single neurons and networks, including synaptic activity, collision of impulses, spectral analysis of neural data, stimulated activity (both deterministic and non-deterministic) of neurons, and the amount of information contained in neural signals. Holden has given a broad summary of some of the models described in the present notes, largely in the form of end results. However, any apparent overlap between these notes and some chapters in Holden's is only superficial. Thus the present work seeks to treat the single neuron model in its own right: only one aspect of neural behaviour, namely, spontaneous activity of single isolated neurons,is modelled in great detail, a critical discussion of the effectiveness of the model being frequently attempted.

CHAPTER 1

SOME BASIC NEUROPHYSIOLOGY

In this chapter, an elementary introduction to neurophysiology is given. The account, though rather oversimplified, serves to make this monograph somewhat self-contained. Standard references (for example: Ochs, 1965; Katz, 1966; Eyzaguirre, 1969) may be consulted for further details.

The long-held view that transmission of signals in the nervous system is due to protoplasmic continuity (i.e., by cells that are fused to one another) broke down after the identification by Ramon y Cajal of single cells called neurons that are both anatomically and physiologically distinct. This led to the neuron doctrine, which, in essence, states that neural signals are for the greater part confined to the cell bodies and the fibers connecting them. The function of the nervous system may then be understood as activity that results from the integration of neurons, a concept introduced by Sherrington. This is the basis of modern neurophysiology. In the following sections, the neuron and its parts are described and their function, rather than structure, emphasised.

1.1 The neuron

There are many kinds of neurons, with a wide range of structural and functional differences. Nevertheless, four basic components can be identified (Figure 1.1.1):

a) the soma or cell body with an outer membrane: this is the central processing element,
b) dendrites, which are hairlike processes emanating from the soma,
c) one or more synapses: these are the input transducers, and
d) the axon: this is the output element and it leads to a neighbouring neuron; it may or may not be myelinated, i.e., covered with a sheath of myelin.

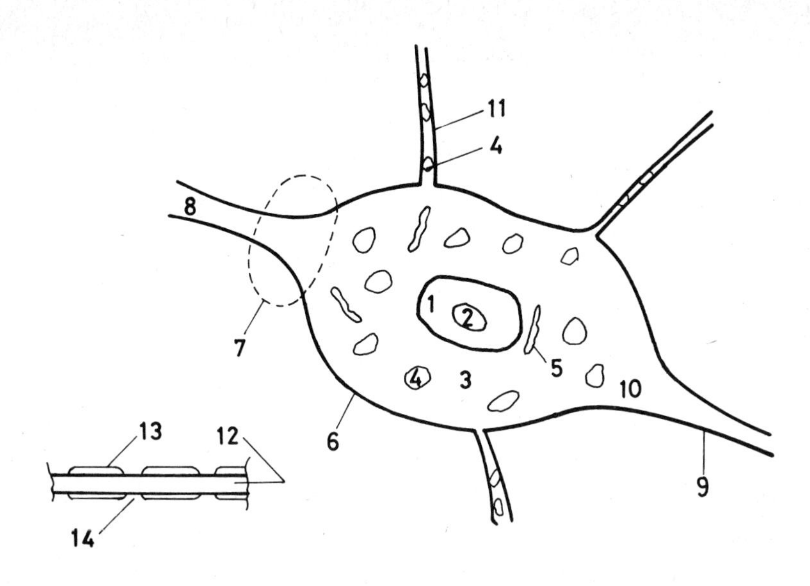

1. Nucleus	2. Nucleolus	3. Soma
4. Nissl body	5. Ribosome	6. Cell membrane
7. Synaptic region (see Fig.1.1.3)	8. Incoming axon	9. Outgoing axon
	10. Axon Hillock	11. Dendrite
12. Myelinated nerve	13. Myelin	14. Node of Ranvier

FIG.1.1.1. THE NEURON (schematic)

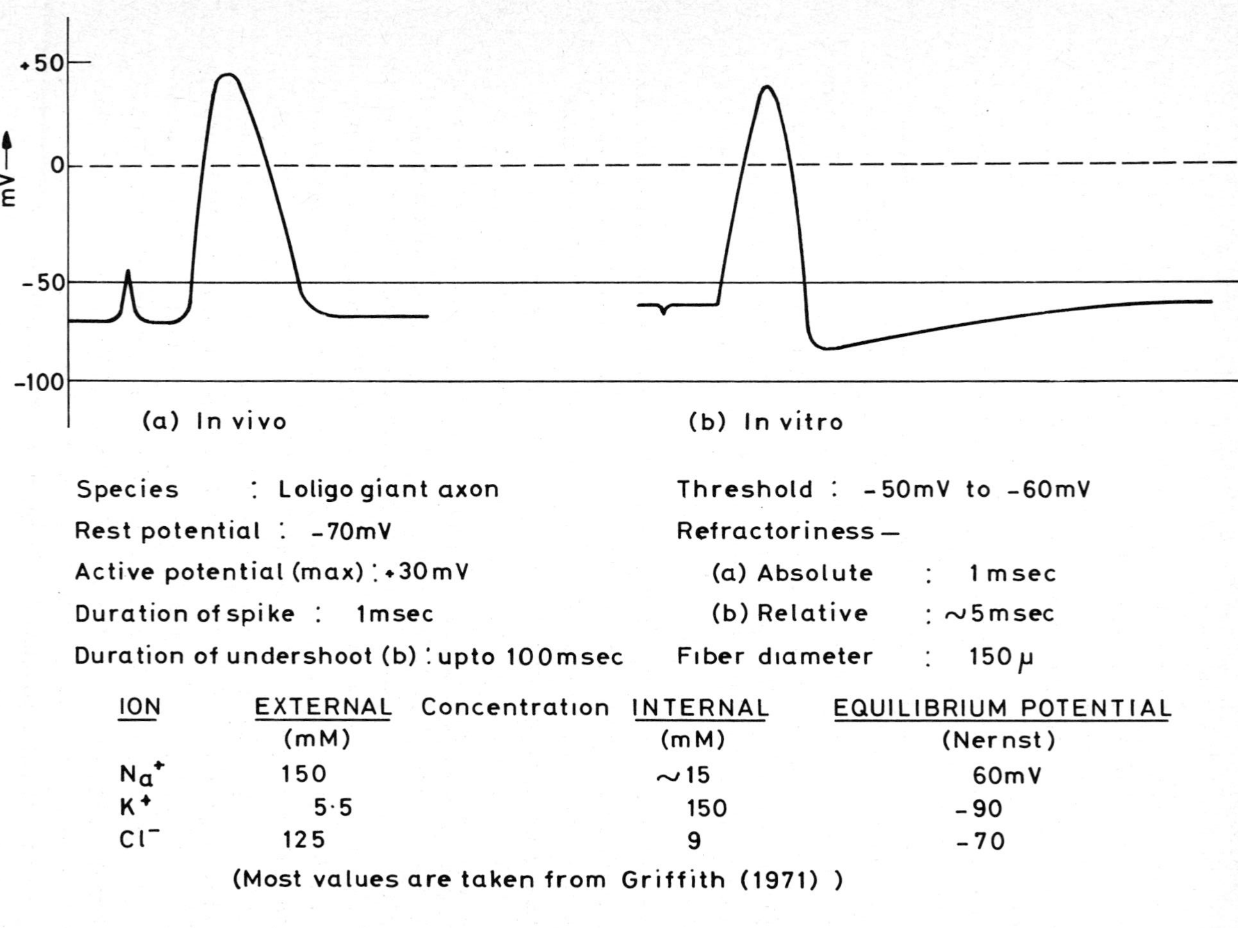

Species : Loligo giant axon

Rest potential : -70mV

Active potential (max) : +30mV

Duration of spike : 1msec

Duration of undershoot (b) : upto 100msec

Threshold : -50mV to -60mV

Refractoriness —

(a) Absolute : 1 msec

(b) Relative : ~5msec

Fiber diameter : 150 μ

ION	EXTERNAL (mM)	Concentration	INTERNAL (mM)	EQUILIBRIUM POTENTIAL (Nernst)
Na^+	150		~15	60mV
K^+	5·5		150	-90
Cl^-	125		9	-70

(Most values are taken from Griffith (1971))

FIG. 1.1.2. THE ACTION POTENTIAL (After Ochs, 1965, p.71)

1.1.1 The axon

In very simple terms, the axon is a cylindrical semi-permeable membrane containing axoplasm and surrounded by extra-cellular fluid. The internal and external fluids are salts, mainly potassium chloride and sodium chloride, that are ionised in solution. The concentration of potassium ions inside is much higher than that outside. This leads to a concentration gradient across the membrane; there is a movement of the excess potassium ions to the outside. This diffusion of ions disturbs the charge balance and results in an electric field which opposes the 'chemical field'. Equilibrium is attained when the two forces are equal, resulting in a potential difference across the membrane, the inside being more negative. This is called the rest potential, and in the squid, Loligo, giant axon is about 70 millivolts. The membrane is said to be 'polarised' at rest.

Equilibrium is disturbed by either changing the extracellular ionic concentration or externally applying an electrical potential gradient across the membrane. If this change is such as to make the inside more negative, it is called hyperpolarisation; if it causes the membrane potential to become more positive, it is called depolarisation. If the depolarisation is large enough to make the membrane potential cross a certain value called the threshold, the potential across the stimulated part shoots up and then returns to the rest level. The peak of this impulse-like change is greater than the threshold value so that adjacent parts of the membrane are excited and the phenomenon propagates along the axon by contiguous stimulation, in general away from the neuron. The waveform has a characteristic shape (Fig.1.1.2), and, since it has only two states, the excited and the unexcited, and none in between, it is described as an all-or-none phenomenon. The above series of events is called the action potential. Hodgkin and Huxley (1952 a,b,c,d), in their classic studies on the Loligo axon, were able to explain many features of the action potential on an unmyelinated nerve fiber. The Hodgkin-Huxley equations provide a neat mathematical model for the action potential, though the physical significance of some of the variables involved is not clear.

In myelinated nerves, a thick sheath of Schwann cells surrounds the fiber, with exposures of 1 micron every millimeter along the length. These breaks in myelination are called nodes of Ranvier

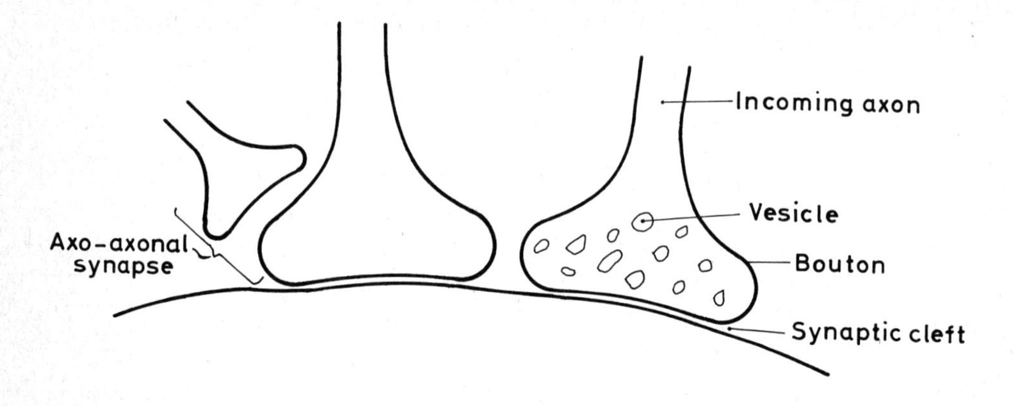

FIG.1.1.3. THE SYNAPSE (schematic)

(see Fig.1.1.1). Unlike in unmyelinated fibers, where the propagation of the action potential is continuous, in myelinated fibers the propagation occurs in jumps from node to node. Except for this, the properties of myelinated fibers are much the same as of unmyelinated fibers (Tasaki, 1968).

When an action potential has been generated, the region of stimulation cannot be excited for a short time thereafter. This property is called refractoriness. Immediately following an action potential, the region cannot for some time generate another, with any strength of stimulus. This is the absolute refractory period and lasts for about 1 millisecond. After this, there is a period of about 3 milliseconds when excitation is possible, but only with a very powerful stimulus: the threshold being very high but not infinite, decreases as the region returns to equilibrium. This is the relative refractory period.

There is a more puzzling property of nerve fibers called accommodation. If the stimulus is applied at a sufficiently slow rate, the fiber is not excited even if the stimulus exceeds the threshold. The fiber seems to 'accommodate'itself to the change. What is more surprising is that with a sufficiently slow rate of rise of the stimulus, the fiber remains unexcited, however high the stimulus may be. The upper limit appears to be only the breakdown potential of the membrane. The process is probably brought about by the elevation of the threshold (Ochs, 1965, p.51); the facts surrounding this are, however, still obscure.

1.1.2 The synapse

Interaction of neurons takes place in the synaptic region (Fig. 1.1.3) by transfer of the activity of one neuron to another by the axon. The terminal end of an axon broadens into a bulge called the bouton adjacent to the cell membrane or a dendrite. Generally, the bouton does not make physical contact with the membrane; a cleft separates the two. Small packets called vesicles are found in the bouton which contain a chemical known as the transmitter, which varies with the kind of junction. This transmitter acts as a mediator in the transfer of activity from the axon to the soma. The arrival of an action potential at the bouton releases the chemical from the vesicles. The transmitter molecules modify the permeability of the membrane to different ions in different ways. If the resulting change in the potential is positive it is called an <u>excitatory post-synaptic potential</u> (EPSP), and if negative an <u>inhibitory post-synaptic potential</u> (IPSP). These changes are quantal in nature; the quantal size may have a probability distribution (Martin and Pilar, 1964; Knight 1972a; Bornstein, 1974). In the absence of an input, the membrane potential tends to decay to the rest level. This decay at subliminal levels (i.e., below threshold) is roughly exponential (Ochs, 1965, p. 45).

There are many kinds of synapses, the differences being in the structure and function in the pre-synaptic and the post-synaptic regions. More than one kind may be found in the same cell. For instance, seven types are found in the hippocampal pyramidal cell and five in the motoneurone (Eccles, 1973, p.8). In the present work only two classes based on

a) whether synaptic transmission is excitatory or inhibitory, and

b) whether the transmission is pre-synaptic or post-synaptic

are considered. The difference in a) is due to different transmitters. Thus acetyl-choline has in general an excitatory effect and gamma aminobutyric acid inhibitory, though there are exceptions to this. The difference in b) is due to a structural difference. When the axon of a neuron synapses directly on the membrane of the soma, the effect of an action potential on the axon is purely post-synaptic. An incoming axon may not, however, synapse directly on the soma but make contact with the synapse of another incoming axon (Ochs, 1965, p. 183; Hirst and McKirdy, 1974). Such a synapse is called an axo-axonal synapse (see Fig.1.1.3). The effect of an action potential at the axo-axonal synapse is to prevent or inhibit the release of transmitter due to an action potential on the directly synapsing axon (Eccles, 1973, pp 91-92). This kind of inhibition is called <u>pre-synaptic inhibition</u>. It is found only in the lower nervous system, for example at the primary afferent level(Eccles, 1973, p.93), and is only infrequently found in the higher nervous system.

It should be noted that whether an impulse causes excitation or inhibition is decided by the chemical transmitter; it cannot be known from the action potential. There is, however, a conjecture known as <u>Dale's principle</u>, which states that the chemical trans-

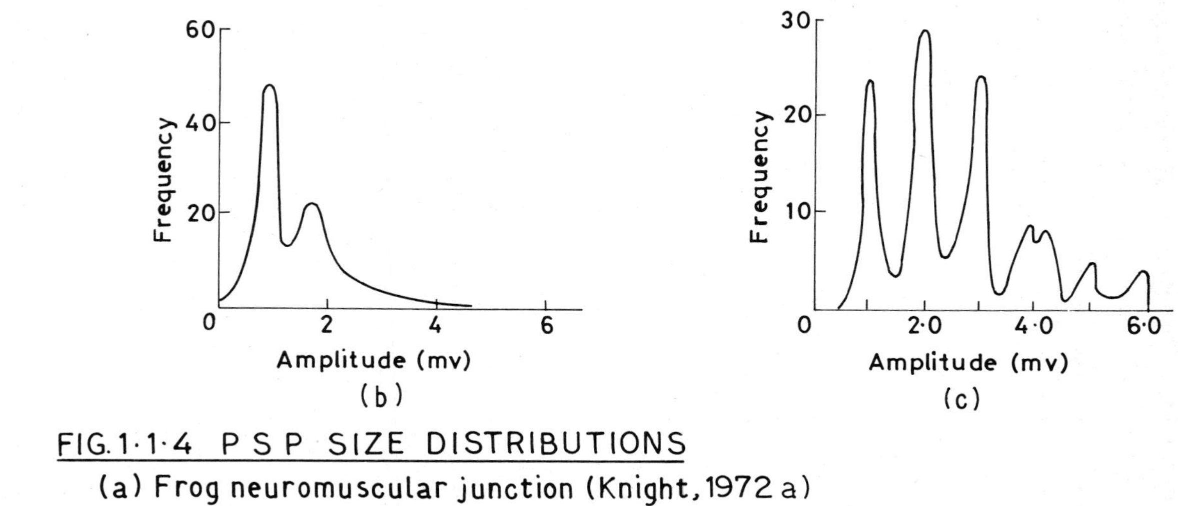

FIG.1·1·4 P S P SIZE DISTRIBUTIONS

(a) Frog neuromuscular junction (Knight, 1972 a)
(b) Ciliary ganglion of chick (Martin and Pilar, 1964)
(c) Pre-ganglion synapse in hypogastric ganglia of male guinea pigs (Bornstein, 1974)

mitter released by the different nerve endings of nerves emanating from the same cell is the same. Thus a neuron can be an excitatory neuron or an inhibitory neuron but not both (Eccles, 1973, p.86).

Synapses which do not involve a chemical mediator but have a direct post-synaptic electrical effect are known (Furshpan, 1964), and are sometimes found along with chemical synapses in the same species. Electrical transmission is used only at special sites in the brains of lower animals, and only rarely in the mammalian brain. The reason for this is the large impedance mismatch between the nerve fiber and the neuron or muscle fiber. Chemical transmission overcomes this difficulty by introducing an amplification of 100 (Eccles, 1973, p. 63).

1.1.3 The soma

Nuclear and cytoplasmic structure and metabolic function are much the same as in most other kinds of cells, except that the neuron does not normally divide. The outputs (i.e., axons) of neighbouring neurons converge on the soma or the dendrites. The number of synapses may vary from just one (Bornstein, 1974) to as many as 80000 (Eccles, 1973, p.100). Thus post-synaptic potentials are induced at several points along the membrane and with successive arrivals of action potentials. These are summed up in the cell. This phenomenon is familiarly known as spatial and temporal summation. The summation is linear at the soma, e.g. if there are three independent sets of excitatory synapses the resulting EPSP is the sum of the three individual EPSPs (Eccles, 1973, p. 71). Similarly with IPSPs; but it is nonlinear in the dendrites. As in the axon, recovery effects keep taking place in the absence of inputs, i.e., the membrane is 'leaky'. When the integrated effect exceeds the threshold of excitation of membrane, the neuron 'fires', sending an action potential along its axon. The origin of the axon at the soma, known as the hillock (see Fig 1.1.1), has a lower threshold than the other parts of the membrane and is thus the impulse generating region of a neuron. Like the axon, the soma also has the properties of refractoriness and accommodation.

1.1.4 Dendrites

These hair-like processes protruding from the soma may be numerous and thus present an increased area of contact to incoming axons, each of which may branch out to form many synapses. The role of dendrites in the firing of the cell is still hazy. Action potentials can arise in dendrites also, but the behaviour of dendrites is highly non-linear and is not understood clearly.

1.2 Types of neurons

There is a wide variety of neurons in the nervous system. They can be roughly divided into

1) sensory neurons,
2) interneurons,
3) motoneurons, and
4) neurons in the central nervous system.

The classification is based only on general functional similarities, and within each class there are neurons with highly specialised functions. For instance, the Renshaw cells are interneurons in the spinal cord which have a special kind of inhibitory effect (Ochs, 1965, pp 333 and 356). Similarly there are several special types of neurons in the central nervous system.

In these notes, interest will center only on the different types of mechanisms of interaction of excitatory and inhibitory impulses and the changes they cause in the neuron membrane potential, though the type of neuron for which a particular model may be suitable is generally indicated.

CHAPTER 2

SIGNALS IN THE NERVOUS SYSTEM

It is evident from the first chapter that information is passed from one neuron to another through an electrochemical change resulting in the propagation of a wave along the connecting fibers. Thus the signal in the nervous system is the travelling action potential. The behaviour of an organism depends on interaction with its environment as well as on changes in its internal milieu, and in an advanced organism thousands of neurons are interconnected. Signal processing in such a network is therefore complex and cannot be described in simple terms. However, there is one basic property of all neural signals — the property that information is coded in a sequence of action potentials. This is true at all levels — whether at the periphery where information is received from the outside world through sensory organs or in the central processing areas of the brain. In the frog, for example, distension of the stretch receptor produces a slow action potential (Ochs, 1965, p. 222) known as the _generator potential_. This is translated into a sequence of firings by the sensory neuron, the rate being proportional to the generator potential level. The sensory message is, therefore, a series of impulses. Similarly, in the central nervous system all signalling is done by sequences of impulses (Eccles, 1973, p.12). On the output side, a sequence of firings of a motoneurone activates a muscle fiber (Ochs, 1965, p.157). Therefore, the neural message is coded in sequences of impulses; there are very few exceptions. It may then be presumed that this message is entirely contained in the rate of firing rather than in the shape of the action potential. In the next section this assumption is shown to be correct from logical argument as well as from experimental records.

2.1 Action potentials as point events — point processes in the nervous system

Rushton (1961) has advanced the following argument in support of the assumption that neural information is transmitted by frequency modulation. Nerve signals have to propagate along axons that are immersed in a lossy medium. Therefore they need to be boosted up all along the fibers. In myelinated fibers, this is done at the nodes of Ranvier. Any loss or gain of amplitude due to imperfect amplification, say an error of 1 %, is multiplied at the hundreds of nodes that sensory information has to pass through before it can reach the spinal cord, resulting in a factor of difference that is as high as 1000. Since there is no evidence of high-tolerance components in the nervous system, amplitude modulation can be ruled out. The alternative is frequency modulation. The impulses transmitted have the same waveform; it is the frequency of discharge that conveys the information. Refractoriness prevents one impulse from running into another. Thus, only the point event of the action potential (i.e., any consistent point during the course of the action potential can be assumed to represent the event of the action potential) is significant. Hence the message may be assumed to be conveyed by a sequence of point events in time. Experimental records tend to confirm this assumption (see Fig.2.1.1). Owing to the spiky appearance of these sequences, they are often called 'spike trains'.

It must be noted here that the assumption that the discharge of a neuron always occurs as a sequence of action potentials approximated by point events is not universally valid. It is not true when considering transmission over short distances, nor in certain species in which transmission is impulseless, e.g. in the crab muscle receptor (Bush and Roberts, 1968) and in reciprocal synapses (Eccles, 1973, p. 97). However, these exceptions are few and the assumption made above is valid for almost the entire nervous system. The models presented in these notes rest entirely on this assumption.

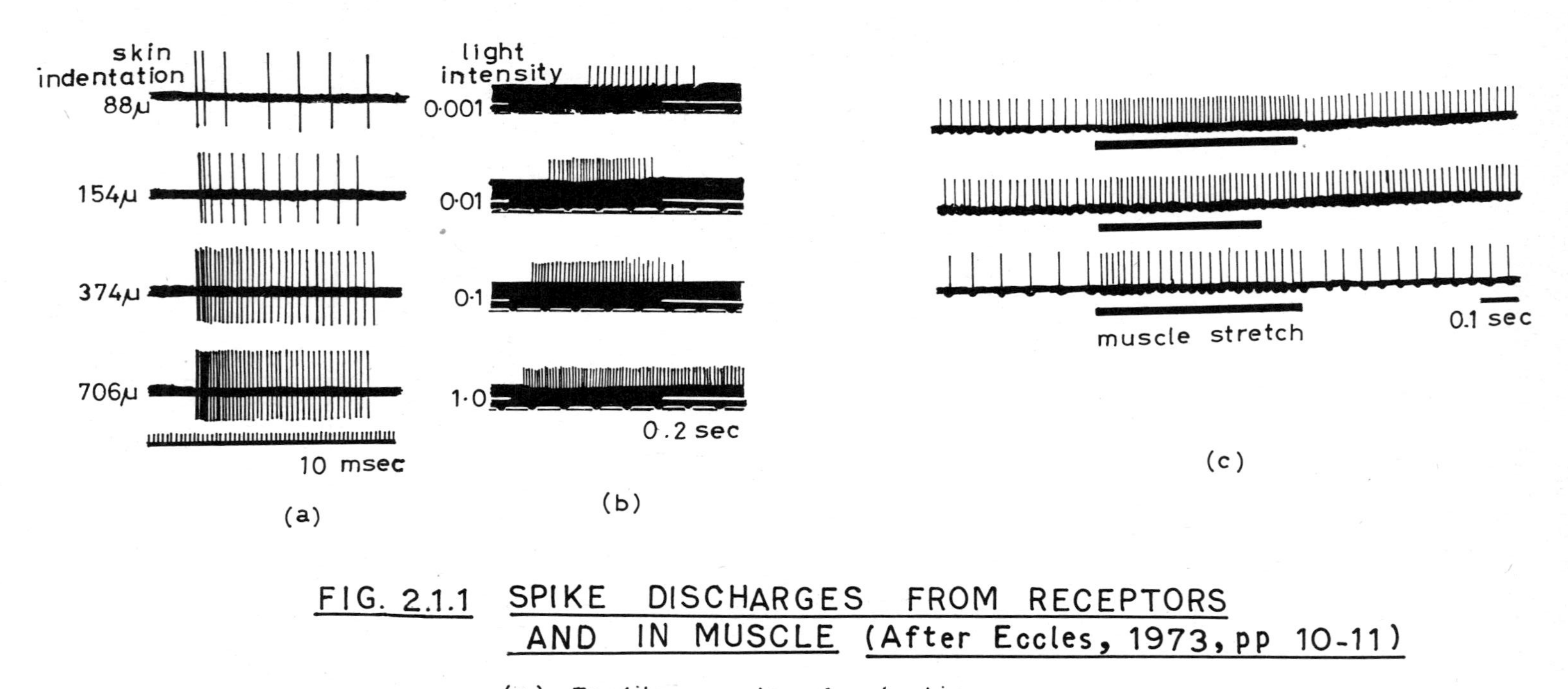

FIG. 2.1.1 SPIKE DISCHARGES FROM RECEPTORS AND IN MUSCLE (After Eccles, 1973, pp 10-11)

(a) Tactile receptor of cat skin

(b) Limulus eye

(c) Annulospiral ending of cat soleus muscle

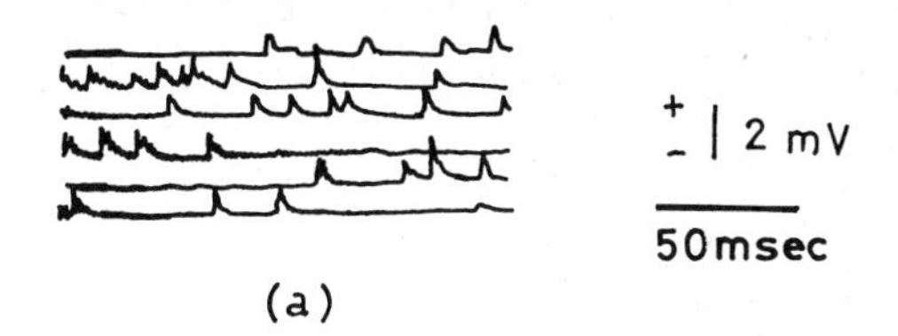

(a)

10 sec

10 mV

(b)

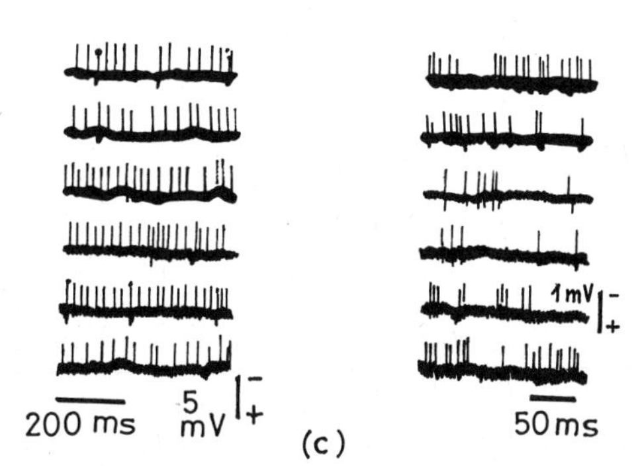

(c)

FIG. 2.2.1 SPONTANEOUS ACTIVITY IN THE NERVOUS SYSTEM.

(After Eccles, 1973, p 45 & 134, Dodge et al., 1968)

(a) Miniature end-plate potentials.

(b) Generator potentials in Limulus cells.

(c) Purkyně and fastigial cells.

2.2 Spontaneous activity in neurons

In the early fifties, Fatt and Katz discovered some unusual behaviour in the neuromuscular junction (del Castillo and Katz, 1956). In the absence of any stimulation, small random depolarising potentials were observed in the end-plate region of frog muscle fibers. These are called miniature end-plate potentials (mepp's) with quantal behaviour occurring as a random release of packets of acetylcholine from the synaptic vesicles, and with a mean rate of 1 to 100 per second. This put an end to the long-held view that peripheral systems are deterministic. Evidence of such random behaviour has been found in many other situations. Levick and Williams (1964) have found in a detailed study a variety of patterns in irregular firing of neurons in the cat lateral geniculate nucleus in the dark. Intracellular recordings made in *Limulus* eye cells show that the generator potential arises out of superposition of fluctuations in the membrane conductance (Dodge etal., 1968). In the brain there is almost always background firing of neurons (Eccles, 1973, p. 101). Some of these records are reproduced in Figure 2.2.1. See also Figure 3.1.1.

Thus neurons are found to fire spontaneously, i.e., they fire even in the absence of external stimuli. The firing occurs at random and hence has the characteristics of a stochastic process. From Rushton's argument (Section 2.1), the spontaneous activity in neurons may be considered a stochastic point process. The stochastic properties of neuron spike trains have been extensively studied and brief reviews have appeared in the literature (Stein, 1972; Knight, 1972a). The firing process is inherently a complicated process owing to the neurons being interconnected; this makes theoretical study difficult. However, it is possible to construct meaningful models of spontaneous activity in single neurons by making certain assumptions. In the next chapter, the basis of modelling is discussed and the mathematical neuron introduced.

CHAPTER 3

STOCHASTIC MODELLING OF SINGLE NEURON SPIKE TRAINS

3.1 Characteristics of a neuron spike train

The firing sequence of a spontaneously active neuron shows a wide variety of patterns. For instance, a 110-minute record of the **mean** rate of firing of single units in the cat lateral geniculate nucleus in the dark (Levick and Williams, 1964) shows a stable mean for 21 minutes followed by irregular transient excitation (25 mins.), stable mean rate (8 mins.), bistable behaviour (27 mins.) and irregular transient firing (29 mins.). There are many statistical techniques for analysing the data, and any mathematical model must be compared with these statistical descriptions, which have varying degrees of complexity. At one end, the entire sequence can be described by listing the lengths of the time interval between successive firings. Such an approach is cumbersome and unrewarding in a search for underlying firing patterns. At the other end the mean rate of firing gives a single number representing the entire firing process and is a drastic condensation of the data. Statistical characteristics between these two extremes are generally used for comparison with the model. One such description is the histogram of the interval between two successive spikes.

If a stochastic process is stationary, a time translation of the origin does not change the statistical properties of the process. Thus, in a point process the interval histogram remains unchanged if a different period of the record is used. If, in addition to being stationary, the process has the intervals between two successive events identically and independently distributed (iid), it is called a renewal process (Cox, 1962). This results in a considerable amount of simplification since the probability density function (pdf) of the intervals is sufficient to describe the process completely.

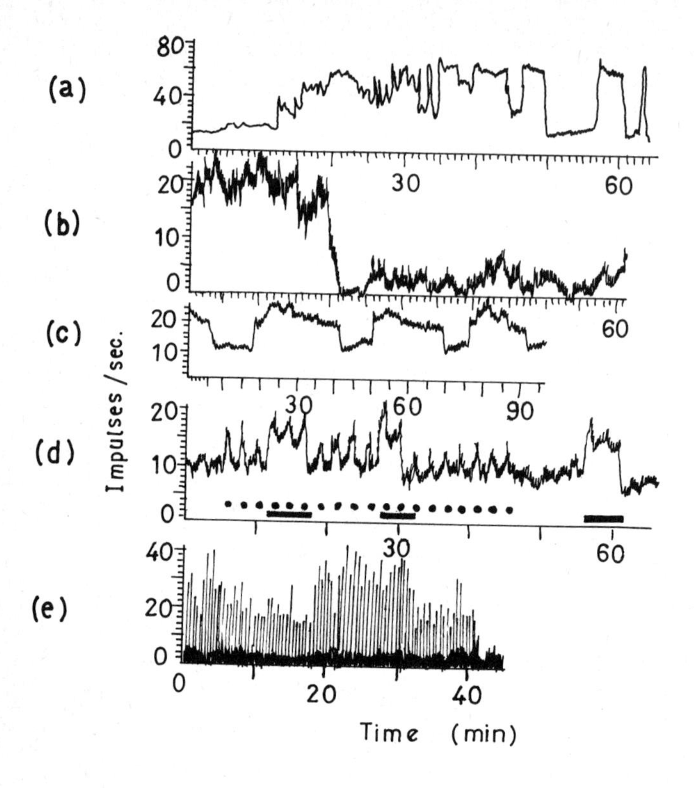

FIG. 3.1.1. DARK FIRING IN CAT LATERAL GENICULATE NUCLEUS (Bishop et al., 1964)

(a), (b) Irregularly unstable

(c) Slow cyclic firing

(d) Two different cyclic patterns superimposed

(e) Fast cyclic firing

Now, it is generally accepted that neuron spike trains are stationary, though such an opinion is not unanimous. Thus, for example, Bishop, Levick and Williams (1964) performed tests for stationarity on their experimental data (Levick and Williams, 1964) and concluded that the sequences are indeed stationary. On the other hand, Stein (1972) states that spike sequences are rarely so even to first order. However, it may be assumed that the statistical properties of spike trains are fairly constant over considerable lengths of time. This not only simplifies the mathematics of modelling but also results in models having practical utility. The second simplification, viz., that the firing sequence is a renewal process, appears to be extremely naive. Surprisingly, however, statistical tests reveal that firing sequences do show such characteristics over long intervals of time. Thus Ekholm (1972) observes that more than two-thirds of recordings of spontaneous activity made by his group exhibit the characteristics of a renewal process. Skvaril, Radil-Weiss, Bohdanecky and Syka (1971) performed statistical analyses on portions of records (stable samples of 5 to 30 mins. with 1000 to 10000 spikes from a single neuron) from several neurons in the mesencephalic reticular formation in rats and the rostral colliculus in cats. Basing their analysis on the interval histogram, mean interspike interval, autocorrelation histogram, shuffled interval autocorrelation histogram, etc., they concluded that the spontaneous activity is completely described by the interval histogram.

There is, therefore, some justification for the use of renewal theory in neuron modelling. Since the renewal process is completely described by the interval pdf, comparison may be made with the experimentally obtained interval histogram. If the theoretical pdf and the histogram show close agreement, the modeller is encouraged to develop his scheme further. It should be noted here that the problem of uniqueness exists: there may be several formulations leading to the same type of interval pdf.

Several shapes have been obtained for the interval histogram by different investigators (Fetz and Gerstein, 1963; Poggio and Viernstein, 1964; Levick and Williams, 1964; Skvaril et al, 1971). These may be classified as

i) exponential (with a small dead time, i.e., refractoriness),

ii) Gaussian,

iii) gamma, usually of order 2,

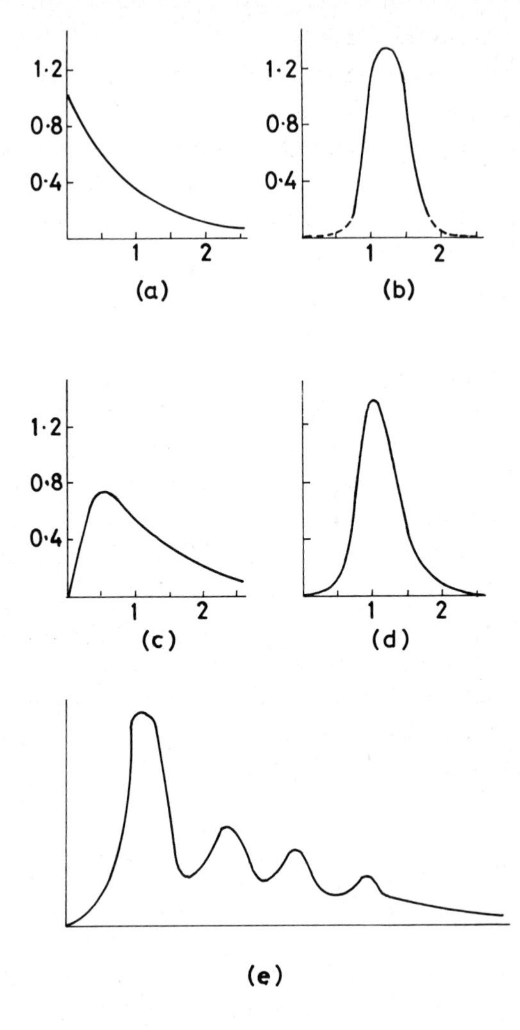

(a) Exponential

(b) Gaussian

(c) Gamma

(d) Symmetrical

(e) Multimodal

FIG.3·1·2 TYPES OF INTERVAL HISTOGRAMS OBTAINED FROM SPONTANEOUS ACTIVITY

iv) symmetrical,
v) long-tailed, and
vi) multimodal.

See Fig.3.1.2. Such variety shows the extent of variability in neurons. A neuron model must have sufficient flexibility to generate many kinds of interval distributions and include as many features of real neurons as possible.

It must be remembered that the renewal model represents only a particular phase of activity: a neuron whose firing sequence is a renewal process during some period of time may lapse into a state of activity that is highly correlated or even periodic. In fact, the renewal property is incompatible with the concept of memory because of the regenerative property of each one of the renewal points. Methods other than those of renewal theory must be used to represent such behaviour. These include frequency domain analysis (Knight, 1972a), the use of coherence functions (Stein, 1972), etc. A general description of the process as a point process and its characterisation through product densities appears to be useful in theoretical formulation of such non-renewal behaviour, especially because it can be used to study non-stationary characteristics as well (see Section 10.7).

3.2 The mathematical neuron

The first mathematical neuron was the McCulloch-Pitts neuron, also known as the logical neuron (Griffith, 1971). This is an extreme idealisation of a real neuron which assumes that the neuron is a binary device which can change its state only at discrete instants of time $t_0 + n\tau$, where n is an integer and τ a constant. Though it incorporates in a simple form the features of threshold, spatial summation and all-or-none output, it is unrealistic in its assumption of impulses arriving at discrete instants of time. It is therefore not useful in modelling spontaneous activity.

In modelling spontaneous firing, the mathematical neuron must essentially take the form of a first-passage time problem (Cox, 1962, p. 97), though there are other formulations that are less realistic (see below). The membrane potential, which has a characteristic resting value, is modified continually by impulses arriving at random intervals of time so that it moves towards the threshold or away from it. When the potential crosses the threshold level, the neuron fires, giving an output spike. Immediately after this, the membrane potential returns to the rest level. A firing is followed by a dead time during which no firing can occur. This process repeats itself: the output of a neuron is the sequence of firings. Due to the stochastic nature of the problem, the sequence is a stochastic process. Viewed at a different angle, the neuron is a filter which acts on several input sequences to give as its output a stream of point events with a reduced frequency. This second approach has been helpful in modelling certain inhibitory mechanisms present in neurons.

Neuron models in the literature can be classified in many ways. Considered purely as mathematical processes, they can be divided into

a) superposition models,
b) deletion models,
c) diffusion models,
d) counter models,
e) discrete state random walk models, and
f) continuous state random walk models.

A more realistic classification is based on the physiological mechanism involved. Thus a neuron is a threshold device which receives as its inputs streams of excitatory and inhibitory impulses. These streams can interact in different ways —

1) a. pre-synaptically, with an impulse with an inhibitory effect suppressing an impulse with an excitatory effect that follows it, or
   b. post-synaptically, with an impulse with an excitatory effect increasing the membrane potential (this resembles the EPSP) and an impulse with an inhibitory effect either decreasing the membrane potential (this resembles the IPSP) or clamping it to the rest level,
2) a. independently of each other, i.e., the time sequence of one train does not depend on that of the other, or
   b. one input process is controlled by the other.

The first classification will be used here because it is more convenient. Nevertheless, the physiological properties of neurons will be emphasised and the mathematical models related to real neurons, as the aim is to model real neurons rather than study mathematical processes. The different types of models are presented in Chapters 4 to 9. The last chapter gives a discussion of some of the physiological features of real neurons in relation to mathematical models.

In what follows, the following features are assumed, unless otherwise stated, for the hypothetical neuron.

1) The neuron is considered isolated from other neurons, i.e., neural interconnections in a network are ignored.
2) There are in general, two input sequences — a sequence of 'excitatories' or 'e-events' and a sequence of 'inhibitories' or 'i-events'. The adjectives 'excitatory' and 'inhibitory' merely refer to the effect that an impulse has in the synaptic region.
3) The terms 'impulse', 'event' and 'spike' are used interchangeably.
4) The input sequences are renewal processes.
5) The change in the membrane potential due to an incoming impulse is instantaneous.
6) Dendrites and their effects are not considered.
7) Spatial summation is ignored, though the effect is usually brought in through superposition of input sequences.
8) There is no synaptic delay (see Section 10.5).

CHAPTER 4

SUPERPOSITION MODELS

Many kinds of neurons have a large number of synapses. Since at each synapse there is an incident impulse sequence, in effect a number of point processes are superposed at the membrane, leading to a composite point process. Neuron models based on such superposition of point processes have been proposed in the literature. In this chapter, some of these models are discussed. Before the models are introduced, the basic theory of superposition of point processes is presented. Section 4.1 deals with renewal processes, and in Section 4.2 the limiting behaviour of a large number of stationary regular point processes that are superposed is considered. The subject of superposition is fairly advanced, but here only some basic concepts and results are presented. The treatment follows for the greater part Cox (1962) and Khintchine (1960). Section 4.3 discusses some models based on superposition.

4.1 Superposition of renewal processes

In Section 3.1, it was stated that renewal theory has been widely used in modelling spontaneous activity in neurons and that there is sufficient experimental evidence to justify this. Consider the hypothetical neuron to have a number of input sequences that are renewal processes. There occurs a superposition of these processes at the membrane. In this section, some results from the mathematical theory of superposition of regular stationary renewal processes are given. Certain conditions are imposed on the primary, i.e., incident,processes, namely, that they are independent of each other and have identical characteristics. In the next section more general processes are considered.

Theorem 4.1.1

If n Poisson processes with parameters $\lambda_1, \lambda_2, \ldots, \lambda_n$ respectively are superposed, the pooled process is a Poisson process with parameter $\lambda = \lambda_1 + \lambda_2 + \ldots + \lambda_n$.

Proof. Consider an arbitrary instant of time in the pooled process. Every instant of time is a regenerative point for the Poisson process. Hence

Prob [time to the first event in the pooled process from the arbitrary time origin > t] =

Prob [time to the first event in the i-th primary process from the arbitrary time origin > t, i = 1,2,...,n] (4.1.1)

$$= \prod_{i=1}^{n} \text{Prob [time to the first event in the i-th primary process} > t]$$

$$= \prod_{i=1}^{n} e^{-\lambda_i t} = e^{-\sum_{i=1}^{n} \lambda_i t}, \qquad (4.1.2)$$

the product arising out of the independence of the n processes. This is a characteristic property of the Poisson process. Thus, the pooled process is Poisson with parameter $\sum_{i=1}^{n} \lambda_i$.

Now, consider a pooled sequence of ordinary renewal processes that are identical and independent. Many properties of the pooled process can be obtained from the generating function of the random variable of the number of events in the pooled process.

Theorem 4.1.2

When n independent and identical ordinary renewal processes are pooled, the probability generating function (p.g.f.) $\underline{G}(t,u)$ of N(t), the number of events in the pooled process in (0,t), is given by

$$\underline{G}(t,u) = [G(t,u)]^n \qquad (4.1.3)$$

where G(t,u) is the pgf of the number of events in (0,t) in a primary process.

Proof

$$\underline{N}(t) = \sum_{i=1}^{n} N^{(i)}(t) \tag{4.1.4}$$

where $N^{(i)}(t)$ is the number of events in $(0,t)$ in the i-th primary process. Now

$$\underline{G}(t,u) = \sum_{r=0}^{\infty} u^r \text{ prob } [\underline{N}(t) = r]$$

$$= \sum_{r=0}^{\infty} u^r \text{ prob } [\sum_{i=1}^{n} N^{(i)}(t) = r] \tag{4.1.5}$$

and

$$G(t,u) = \sum_{r=0}^{\infty} u^r \text{ prob } [N^{(i)}(t) = r]. \tag{4.1.6}$$

Using the convolution property,

$$\underline{G}(t,u) = [G(t,u)]^n. \tag{4.1.7}$$

From this, the properties of $\underline{N}(t)$ can be readily obtained. For example

prob $[\underline{N}(t) = i]$ = coefficient of u^i in the expansion of $\underline{G}(t,u)$. (4.1.8)

Similarly, the mean time upto the j-th event in the pooled process can be obtained through $\underline{G}(t,u)$.

Theorem 4.1.3

The mean of the time $\underline{S}_j$ upto the j-th event in the pooled process is given by

$$E(\underline{S}_j) = \text{coefficient of } u^i \text{ in } \frac{u}{1-u} \int_0^{\infty} [G(t,u)]^n dt. \tag{4.1.9}$$

Proof

Now,

$$\text{prob } [\underline{S}_j > t] \quad = \quad \text{prob } [\underline{N}(t) < j] \tag{4.1.10}$$

so that

$$E(\underline{S}_j) \quad = \quad \int_0^\infty t \text{ prob } (\underline{S}_j = t) \, dt$$

$$= \quad \int_0^\infty \text{prob } (S_j > t) \, dt$$

$$= \quad \int_0^\infty \text{prob } (\underline{N}(t) < j) \, dt. \tag{4.1.11}$$

$$\sum_{j=1}^\infty u^j \text{ prob } (\underline{S}_j > t) \quad = \quad \sum_{j=1}^\infty u^j \text{ prob } (\underline{N}(t) < j)$$

$$= \quad \sum_{j=0}^\infty \text{prob } (\underline{N}(t) = j) \left\{ u^{j+1} + u^{j+2} + \cdots \right\}$$

$$= \quad \frac{u}{1-u} [G(t,u)]^n, \tag{4.1.12}$$

using Theorem 4.1.2.
Thus

$$\sum_{j=1}^\infty u^j \int_0^\infty \text{prob } (\underline{S}_j > t) \, dt \quad = \quad \frac{u}{1-u} \int_0^\infty [G(t,u)]^n dt, \tag{4.1.13}$$

that is,

$$\sum_{j=1}^\infty u^j E(\underline{S}_j) \quad = \quad \frac{u}{1-u} \int_0^\infty [G(t,u)]^n \, dt, \tag{4.1.14}$$

from which (4.1.9) follows.

The counting statistics of the pooled process are thus easily obtained from the generating function. In neuron modelling, however, the pdf of the interval between two successive events in the pooled process is more useful. This is because the pdf can be compared with interval histograms obtained from experimental records of spontaneous activity. The following theorem gives an expression for the interval pdf in a pooled process arising from n primary processes that are considered remote from their time origin, i.e., they are equilibrium renewal processes.

Theorem 4.1.4

The pdf $g_X(x)$ of the interval between two successive events in a pooled process obtained by superposing n equilibrium renewal processes that are identical and independent with interval pdf $f(\cdot)$ is given by

$$g_X(x) = \frac{-d}{dx}\left[\bar{F}(x)\left\{\int_x^\infty \frac{\bar{F}(u)}{\mu}\,du\right\}^{n-1}\right] \tag{4.1.15}$$

where

$$\bar{F}(x) = \int_x^\infty f(x)\,dx \tag{4.1.16}$$

and

$$\mu = \int_0^\infty x\,f(x)\,dx. \tag{4.1.17}$$

Proof

Considering n superposed identical equilibrium renewal processes, the mean interval between successive impulses in the pooled process is μ/n. Let U be the backward recurrence time in the pooled process.

$$U = \min(U_1, \cdots, U_n), \tag{4.1.18}$$

where U_i is the backward recurrence time in the i-th primary process. Owing to the primary processes being independent,

$$\begin{aligned} \text{prob}(U > x) &= \text{prob}[\min(U_1, \cdots, U_n) > x] \\ &= \prod_{i=1}^{n} \text{prob}(U_i > x) \\ &= \left\{\int_x^\infty \frac{\bar{F}(u)}{\mu}\,du\right\}^n. \end{aligned} \tag{4.1.19}$$

The pdf of U is

$$\frac{-d}{dx}\,[\text{prob}\ (U > x)] = \frac{n}{\mu}\,\bar{F}(x)\left\{\int_x^\infty \frac{\bar{F}(u)}{\mu}\,du\right\}^{n-1}. \qquad (4.1.20)$$

Consider the survivor function $\bar{G}_X(x)$ of the interval **between** two successive impulses in the pooled process. Then $\bar{G}_X(x)$ is related to U by the following equation:

$$\frac{\bar{G}_X(x)}{\text{mean interval in the pooled process}} = \text{pdf of } U, \qquad (4.1.21)$$

that is,

$$\frac{\bar{G}_X(x)}{\frac{\mu}{n}} = \frac{n}{\mu}\,\bar{F}(x)\left\{\int_x^\infty \frac{\bar{F}(u)}{\mu}\,du\right\}^{n-1}. \qquad (4.1.22)$$

On differentiating $-\bar{G}_X(x)$ with respect to x,

$$g_X(x) = \frac{-d}{dx}\left[\bar{F}(x)\left\{\int_x^\infty \frac{\bar{F}(u)}{\mu}\,du\right\}^{n-1}\right] \qquad (4.1.23)$$

and the proof is complete.

Two examples are considered below.

1) Let $f(x) = \frac{1}{\mu}e^{-\frac{x}{\mu}}$. The primary processes are Poisson.

Then

$$g_X(x) = \frac{n}{\mu}e^{-\frac{nx}{\mu}}. \qquad (4.1.24)$$

The pooled process also is Poisson. This result follows otherwise from Theorem 4.1.1.

2) $f(x) = \frac{x}{2\mu}, \quad 0 \leqslant x \leqslant 2\mu$

$$= 0, \quad x > 2\mu. \qquad (4.1.25)$$

Then

$$\bar{F}(x) = 1 - \frac{x}{2\mu}, \qquad 0 \leqslant x \leqslant 2\mu$$
$$= 0 \quad , \qquad x > 2\mu. \tag{4.1.26}$$

$$g_X(x) = \left(\frac{2n-1}{2\mu}\right)\left(1 - \frac{x}{2\mu}\right)^{2n-2}, \quad 0 \leqslant x \leqslant 2\mu. \tag{4.1.27}$$

Make the transformation

$$Y = n X. \tag{4.1.28}$$

Then the pdf of the new random variable representing the interval between two successive events is given by

$$g_Y(y) = \frac{1}{\mu}\left(1 - \frac{1}{2n}\right)\left(1 - \frac{y}{\mu}\frac{1}{2n}\right)^{2n-2} \tag{4.1.29}$$

Considering the behaviour of $g_Y(y)$ as $n \to \infty$,

$$g_Y(y) \to \frac{1}{\mu}\left(1 - \frac{1}{2n}\right)\left(1 - \frac{y}{\mu}\frac{1}{2n}\right)^{2n}$$

$$\to \frac{1}{\mu} e^{-\frac{y}{\mu}}. \tag{4.1.30}$$

(In the above derivation it has been assumed that the impulse rate in the pooled process

$$E(X) \to 0 \quad \text{as} \quad n \to \infty \tag{4.1.31}$$

such that

$$E(Y) = n E(X) \tag{4.1.32}$$

is finite).

Thus the interval distribution tends to the exponential (with mean $\frac{1}{\mu}$). This is an interesting result, for it suggests that the pooling of a large number of independent identical renewal processes leads to a process that has the intervals between the successive events exponentially distributed. It is now shown that whatever be $f(x)$ the local behaviour of the pooled process when n is large is approximately that of a Poisson process. Here 'local behaviour' means behaviour over periods of time short in relation to the interval between successive events.

__Theorem 4.1.5__ When n identical independent regular equilibrium renewal processes are pooled, the survivor function of the interval between two successive events in the pooled process tends to an exponential function as n tends to ∞.

__Proof:__ Let X be the interval between two successive events in the pooled sequence. Normalising the random variable as

$$Z = \frac{X}{E(X)}, \qquad (4.1.33)$$

the survivor function of Z is, using equation (4.1.22),

$$\bar{G}_Z(x) = \bar{F}\left(\frac{x\mu}{n}\right) \left\{1 - \int_0^{\frac{x\mu}{n}} \frac{\bar{F}(u)}{\mu} \, du \right\}^{n-1} \qquad (4.1.34)$$

from which

$$\lim_{n \to \infty} \bar{G}_Z(x) = e^{-x} \qquad (4.1.35)$$

using the fact that

$$\bar{F}(0) = 1. \qquad (4.1.36)$$

The above result is generalised in the following theorem which is given without proof.

__Theorem 4.1.6__ The asymptotic distribution of intervals in a pooled sequence made up of independent equilibrium renewal processes not necessarily identical whose survivor functions $\bar{F}_i(x)$ are bounded by survivor functions $\bar{A}(x)$ and $\bar{B}(x)$ (i.e., $\bar{A}(x) \leqslant \bar{F}_i(x) \leqslant \bar{B}(x)$) is exponential as $n \to \infty$. When the individual processes are also identical, successive intervals in the pooled process are, in addition, independently and identically distributed (exponentially).

The above results show that under certain conditions the superposition of a large number of renewal processes leads to a process that is locally Poisson. Notice that in the preceding theorems the renewal processes are assumed to be equilibrium processes. This requires that the processes be considered far from their time origin. In Section 3.1, it has already been stated that for purposes of modelling the neuron firing sequence is assumed to be stationary over long intervals of time. The firing sequences may therefore be considered equilibrium processes and hence the above results apply to them. In the next section the asymptotic behaviour discussed above for renewal processes is shown to be true in a more general case, viz., that of superposition of a large number of stationary point processes.

4.2 Superposition of stationary point processes - limiting behaviour

Most renewal models of neuron spike trains assume that the input sequences are renewal processes. However, the output sequence need not be a renewal process, as will become evident in the forthcoming chapters. In most models not more than two input sequences are considered, one being a stream of e-events and the other a stream of i-events. Without having to resort to the very strong assumption that the input sequences are renewal processes, it is possible to construct models using stationary non-renewal input sequences. Consider the superposition of several e-sequences (arriving at the several synapses); the result is a single pooled e-sequence. The primary sequences are from other neurons and need not be renewal processes. Hence the study of superposition of non-renewal processes acquires relevance. The mathematical theory of superposition of stationary point processes has its origin in the study of telephone traffic at an exchange where there are a number of streams (generally independent of each other) of calls converging on the exchange. This latter phenomenon has been studied in detail by Palm, and by Khintchine, who was motivated by problems in queuing theory. Palm defined a function characterising the arrival pattern of calls and used this function to study the superposition of point processes. He showed that, in the limit, superposition of a large number of stationary point processes leads to a Poisson process. Khintchine (1960) removed some of the mathematical defects in Palm's theory and extended his results.

This section is concerned not with the larger theory of superposition but only with the asymptotic behaviour of a pooled sequence as the motive is only to supply the justification for assuming that one or both of the input sequences in the models to be discussed in later chapters are Poisson processes. The theorems are given without proof; the reader is referred to Khintchine's monograph for the details.

4.2.1 Palm functions

Palm defined the following function for a stationary point process:

$$\varphi_{(0)}(t) = \text{Prob}\,[N(t_0,\, t_0 + t) = 0 \,|\, N(t_0 - \Delta\,, t_0) = 1] \tag{4.2.1}$$

i.e., $\varphi_0(t)$ is the probability of no event occurring in the interval $(t_0,\, t_0 + t)$ conditional on the occurrence of an event in $(t_0 - \Delta\,,\, t_0)$. Khintchine went further and defined a whole sequence of functions $\varphi_{(k)}(\cdot)$, $k = 0,1,\cdots$, based on Palm's definition.

$$\varphi_{(k)}(t) = \text{Prob}\,[N(t_0, t_0 + t) = k \,|\, N(t_0 - \Delta\,,\, t_0) = 1] \tag{4.2.2}$$

i.e., $\varphi_{(k)}$ is the probability of k events $(k = 0,1,\ldots)$ occurring in $(t_0, t_0 + t)$ conditional on an event occurring in $(t_0 - \Delta\,, t_0)$. These functions $\varphi_{(k)}(\cdot)$ are called the _Palm functions_. They are related to the unconditional functions

$$v_{(k)}(t) = \text{Prob}\,[N(t_0, t_0 + t) = k] \tag{4.2.3}$$

(i.e., the Palm functions with the conditioning at the origin removed) by the following theorem.

Theorem 4.2.1 The unconditional probability $v_{(k)}(t)$ of k $(k = 0,1,2\ldots)$ events of a stationary point process occurring in an interval $(t_0, t_0 + t)$ is given by

$$v_{(0)}(t) = 1 - \lambda \int_0^t \varphi_{(0)}(u)\, du \tag{4.2.4}$$

and

$$v_{(k)}(t) = \lambda \int_0^t [\varphi_{(k-1)}(u) - \varphi_{(k)}(u)]\, du,$$
$$(k = 1,2,\dots), \qquad (4.2.5)$$

where λ is the stationary value of the first order product density of the point process (see Srinivasan (1974a) for a discussion of product densities).

In the next section the result of superposing a large number of independent stationary point processes is discussed.

4.2.2 Asymptotic behaviour of n stationary point processes superposed

Using the function defined by equation (4.2.1), Palm has shown that, under broad assumptions, the superposition of n stationary sequences leads to a composite sequence in which the intervals between successive events has a survivor function that is exponential, as $n \to \infty$. Consider the superposition of n stationary, regular and mutually independent sequences. For the i-th stream $(i = 1,2,\cdots,n)$ let λ_i be the stationary value, $\varphi_{(0i)}(\cdot)$ its first Palm function (see equation (4.2.1)) and $v_{(ki)}(\cdot)$ the counting function defined by (4.2.5). Let the same quantities for the pooled sequence be denoted by $\underline{\lambda}$, $\underline{\varphi}_{(0)}(\cdot)$ and $\underline{v}_{(k)}(\cdot)$ respectively. The following assumptions are made:

1) As $n \to \infty$, $\lambda_i \to 0$ and $\underline{\lambda}$ $(= \sum_{i=1}^{n} \lambda_i)$ remains constant, so that for any $\varepsilon > 0$, $\lambda_i < \varepsilon$ with n sufficiently large.
2) As $n \to \infty$, with any constant $t > 0$, $\varphi_{(0i)}(t) \to 1$ so that for any $\varepsilon > 0$, $1 - \varphi_{(0i)}(t) < \varepsilon$, n being sufficiently large.

In the neural context, the first assumption means that as the number of input sequences increases, the spike rate of each sequence decreases such that the spike rate of the pooled sequence is bounded. This reflects only the non-local behaviour of the inputs, i.e., of inputs from neurons located afar. Local inputs have a much higher frequency and hence assumption 1) is not valid for them. They have therefore to be separated from the superposition scheme and possibly considered as a bunching or burst phenomenon. The second assumption is interpreted as the absence of bursts in any input sequence. [In this monograph models of burst activity in neurons are not discussed. For a review of such models the reader is referred to a recent survey article by Fienberg (1974)].

Theorem 4.2.2 (Palm's theorem) For a constant $t > 0$, as $n \rightarrow \infty$

$$\underline{v}_{(0)}(t) \rightarrow e^{-\lambda t}. \qquad (4.2.6)$$

Palm assumed that this is a sufficient condition for the pooled process to tend to a Poisson process. Khintchine remarks that this is not so and has given a sufficient condition. His result can be stated as the Palm - Khintchine theorem.

Theorem 4.2.3 (Palm-Khintchine theorem) For a constant $t > 0$, as $n \rightarrow \infty$

$$\underline{v}_{(k)}(t) \rightarrow \frac{e^{-\lambda t}(\lambda t)^k}{k!}. \qquad (4.2.7)$$

For purposes of neural modelling, it is sufficient to note that if n independent stationary point processes are superposed, the asymptotic behaviour of the superposed process is that of a Poisson process. Thus when a neuron has a sufficiently large number of synapses, the input sequences are superposed resulting in a pooled process that is approximately Poisson. This not only leads to mathematically tractable models but also allows simplification of models formulated under more general assumptions.

4.3 Superposition models of neuron spike trains

In superposition models, the output of a neuron consists of the superposed sequence of all the input trains. In making such a supposition it is assumed that every input excitatory impulse fires the neuron and that there is no inhibition. There exist some neurons in the anteroventral cochlear nucleus that have about three cochlear fibers each and it is believed that every input impulse to these neurons is able to fire the neuron (Molnar and

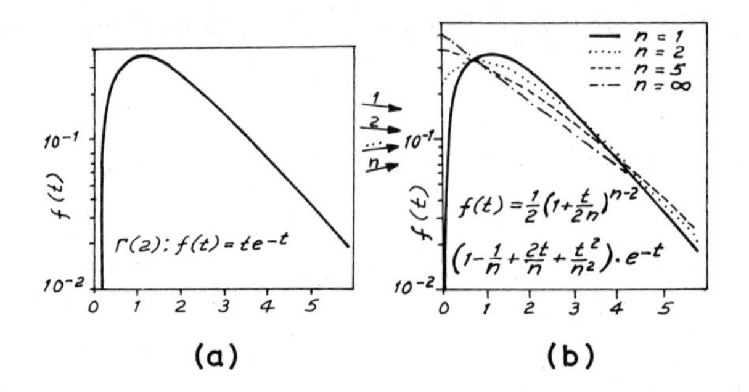

(a) Primary sequence - 2nd order gamma distribution

(b) Superposition of n sequences

FIG. 4.3.1 SUPERPOSITION OF SPIKE TRAINS (Bishop et al 1964)

Pfeiffer, 1968). Superposition models have, therefore, some validity.

4.3.1 Model 4.1 (Bishop, Levick and Williams, 1964)

Bishop, Levick and Williams (1964) proposed a model of neuronal firings assuming a superposition of impulse sequences arriving on several excitatory fibers from the retina converging on a geniculate neuron, each input impulse discharging the neuron so that the output of the cell is the superposed train of all input trains. They made the following assumptions:

1) the interval pdfs of the spike trains generated by the spontaneously active retinal ganglion cells are gamma distributions,
2) refractoriness is not present,
3) the brief high frequency bursts in the spike trains are single point events,
4) there are no inputs to the geniculate nucleus other than those from the retina,
5) there is no inhibition.

Assuming that the input sequences are independent renewal processes, the interval pdf in the pooled process can be found. Bishop et al. considered the specific example

$$f(t) = \frac{k^a t^{a-1} e^{-kt}}{\Gamma(a)} \qquad (4.3.1)$$

with $a = 2$ corresponding to the value obtained by Kuffler et al. (1957) for retinal ganglion cell discharges. This distribution is shown in Fig. 4.3.1. When n streams are superposed the resulting interval pdf of the pooled process has the shape given in Fig. 4.3.1 for different values of n. For comparison purposes k is set equal to $\frac{1}{n}$. (This keeps the mean of the pooled process constant for different n). With $n = 1$, the graph starts from the origin. As n increases the starting point of the graph moves upward on the y-axis. Bishop et al. point out that the shape of the interval pdf approaches that of the short interval histogram obtained in experiments on single units in the cat lateral geniculate nucleus firing in the dark (Levick and Williams, 1964). If a narrow distribution of very short intervals is added to the interval pdf of the pooled process to account for the bursts in the output of retinal ganglia (see assumption (3)), then the theoretical pdf agrees well with the experimental histogram. As n becomes very large, the pdf approaches the exponential (see Fig. 4.3.1). There is a simple relation between the height of the intercept with the y-axis and the number of inputs:

$$\frac{\text{y-intercept with } n \text{ inputs}}{\text{y-intercept with } \infty \text{ inputs}} = 1 - \frac{1}{n} . \qquad (4.3.2)$$

The practical utility of this relation is discussed in Section 10.6.

4.3.2 Model 4.2 A superposition model with two input channels (Ten Hoopen, 1967)

Ten Hoopen (1967) has considered the following model. A neuron has two input channels. An input on either channel gives rise to an output spike. Hence the output of the neuron is the superposition of the two input channels. On one channel the interval pdf is a unimodal distribution. On the other it is exponential or near exponential. The latter assumption is made to account for the short intervals in experimental records of firing of single units in the thalamus. Ten Hoopen has shown that the results of this model agree with the results of statistical analysis of experimental records in a number of cases. The reader is referred to the original paper for a detailed discussion.

4.3.3 Model 4.3 (Sabah and Murphy, 1971)

Sabah and Murphy (1971) proposed a superposition model for the spontaneous activity of cerebellar Purkinje cells. The spike output of ten such cells was analysed and their interspike intervals and variance time curves were studied. Cerebellar Purkinje cells discharge spontaneously with mean frequencies around 10 to 70/sec. Interspike interval distributions of nearly 80% of 116 Purkinje cells of decerebrate cats and lightly anesthetised cats were found to be exponential or with longer tails. The remaining cells gave distributions that are broad and irregular. Sabah and Murphy suggested that the spike output is due to the superposition of spike trains that emanate from the dendritic expanse. There is evidence of at least the larger dendrites in Purkinje cells being able to support spike conduction, which probably arises from a lower threshold. The dendrites are invaded antidromically by action potentials from the soma. Due to the lower threshold the dendrites themselves initiate spikes and the neuron output consists of the pooled sequence of spikes from all the zones. However due to refractoriness not all the dendritic spikes appear at the output. In their report, Sabah and Murphy have estimated 1) the fraction of spikes lost, 2) the number of firing zones from the properties of the pooled sequence (the number of component processes = the number of zones).

4.4 Discussion

For all its success in obtaining interval pdfs that resemble experimental histograms very closely, the superposition model is unsatisfactory.

1) The output spike train in a superposition model has a higher spike rate than the inputs and this is in conflict with most experimental results in which the output frequency is generally much smaller than the input frequency.

2) Even if the input spike trains are assumed to have the refractory property, pooling of sequences destroys this property. If a dead time is included in the pooled sequence, the analysis is complicated.

3) Observed properties like temporal summation of the membrane potential, decay in the absence of inputs, threshold behaviour, etc., are ignored. The only way in which every input impulse can give rise to a discharge is the EPSP being always larger than the threshold. This does occur in some neurons, as stated at the beginning of this section. However most experimental records show a high variability of neuron discharges which cannot be explained by superposition alone. Further, the post-synaptic quantal potentials have been found to have a random size and this must be considered in any realistic model.

4) Inhibitory inputs are ignored completely.

Nevertheless superposition of sequences is an indirect way of bringing in the spatial summation property of neurons. Thus superposition is to be considered a part of a more detailed description of neuron firings rather than the basis of the model, except where there is sufficient evidence to support this belief.

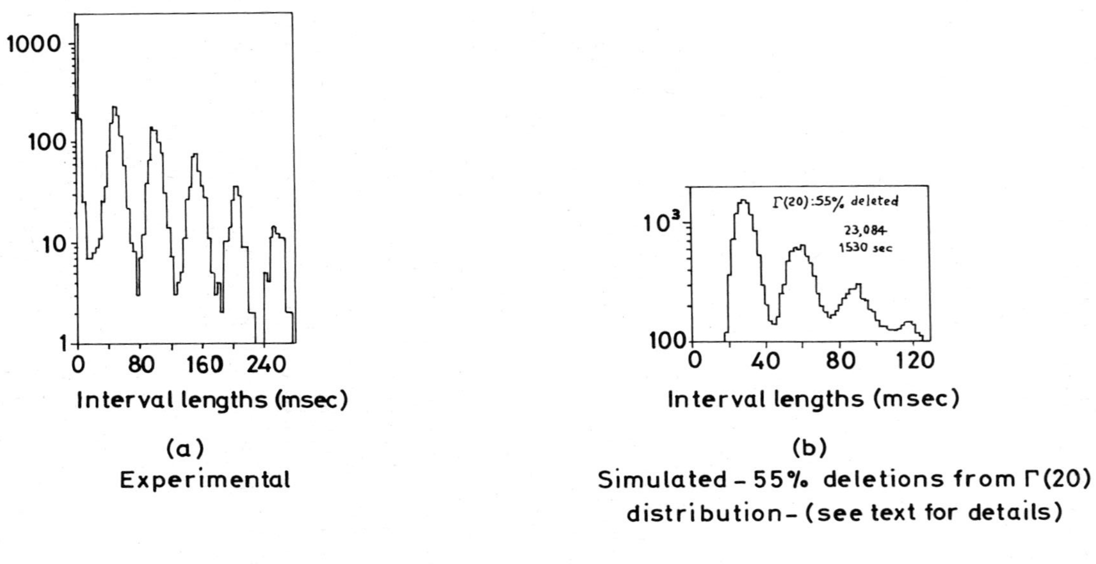

(a)
Experimental

(b)
Simulated – 55% deletions from Γ(20) distribution – (see text for details)

FIG. 5·1 MULTIMODAL DISTRIBUTIONS (Bishop et al., 1964)

CHAPTER 5

DELETION MODELS

In Section 3.1, mention was made of the experiments of Levick and Williams (1964) on the cat lateral geniculate nucleus. Following this, Bishop, Levick and Williams (1964) made a statistical analysis of the data and divided the interval histograms into long-interval histograms and short-interval histograms. The latter show interesting multimodal characteristics (Figure 5.1) with the intervals corresponding to the peaks forming an arithmetic progression. The authors proposed the following model. The neuron has both excitatory and inhibitory fibers incident on it. An impulse on an inhibitory fiber blocks an impulse on an excitatory fiber. Unblocked excitatory inputs are assumed to form the neuron output spike train. Inhibition of this kind has been discussed in Section 1.1.2 as pre-synaptic inhibition, in which action potentials arriving at an axo-axonal synapse inhibit release of excitatory transmitter by an impulse on the direct synapse. This interaction was simulated by Bishop et al. using a radioactive source and an m-ary counter. Thus a sequence whose intervals have a gamma distribution of order m was obtained. This was then gated by a randomly triggered monostable multivibrator, the latter's gating action playing the role of the inhibitory impulse. It was found that the experimentally obtained histograms with their deep troughs can be reproduced if m is about 20.

These findings have resulted in a large number of stochastic models which are generally known as 'deletion models' because an inhibitory impulse 'deletes' or 'pre-inhibits' an excitatory impulse that follows it. (In this chapter the words 'delete' and 'pre-inhibit' are used interchangeably). There are two types of 'deletion models':

1) independent interaction models, in which the two sequences - excitatory and inhibitory - are independent of each other, and

2) dependent interaction models, in which one of the sequences is controlled by the other.

These are discussed in the following sections.

5.1 Deletion models with independent interaction of excitatory and inhibitory sequences

Starting with the above scheme proposed by Bishop et al., Ten Hoopen and Reuver (1965b) proposed a stochastic model of interacting e-events (excitatories) and i-events (inhibitories) in which the two sequences are independent renewal processes with pdf's $\varphi(\cdot)$ and $\psi(\cdot)$ respectively, with an i-event pre-inhibiting the first e-event that follows it. Uninhibited e-events were called 'responses' (or r-events, for convenience), and an expression for the pdf of the interval between two successive r-events was derived. The special cases when $\psi(\cdot)$ or $\varphi(\cdot)$ is exponential were obtained, the former giving multimodal distributions when $\varphi(\cdot)$ is a gamma distribution of order 20, and the latter giving a distribution with a bias towards the origin. This second kind of distribution is to be expected because $\varphi(\cdot)$ being exponential has a bias towards the origin and an r-event coincides with an e-event. Both these cases have been further discussed by Ten Hoopen (1966a) who has studied the interval distributions of the r-sequence generated in the two cases for different distributions of $\varphi(\cdot)$ and $\psi(\cdot)$. Further extensions to this model have been proposed by many authors. Thus

i) an i-event is capable of inhibiting more than one e-event following it,

ii) an i-event is effective only for a time T which can be a constant or a random variable.

Of these, only models with property (i) will be discussed in this chapter. Those with property (ii) are dealt with in Chapter 7 (Counter Models) because of the similarity that the 'lifetime' of an i-event has with dead time in a counter.

5.1.1 Model 5.1 - The basic deletion model (Ten Hoopen and Reuver, 1965b)

The basic deletion model has the following properties:

1) The neuron has two sequences of inputs - excitatory and inhibitory - incident on it.
2) The e-events (excitatories) and the i-events (inhibitories) form independent renewal processes with the pdf of the interval between two successive events being $\varphi(\cdot)$ and $\psi(\cdot)$ respectively.
3) An i-event pre-inhibits the first e-event that follows it.
4) Uninhibited e-events (r-events) constitute the neuron output discharge.

When $\varphi(\cdot)$ and $\psi(\cdot)$ are general functions, an expression for the pdf of the interval between two successive r-events $f(\cdot)$ can be derived in terms of a set of subsidiary variables. When $\psi(\cdot)$ is exponential, the r-events form a renewal process and $f(\cdot)$ characterises the r-sequence completely. When $\psi(\cdot)$ is not exponential, higher order properties are necessary to describe the process.

Lemma 5.1.1 In Model 5.1, if

$$\pi(t) = \lim_{\Delta,\Delta' \to 0} \text{Prob}\,[\text{ the first i-event following an r-event occurs in } (t,t+\Delta') \mid \text{the r-event occurs in } (-\Delta,0)\,]/\Delta' , \qquad (5.1.1)$$

then $\pi(t)$ is given by

$$\pi(t) = \gamma \int_0^\infty \psi(t+w)dw \int_0^w h_e(u)\,du \int_{w-u}^\infty \varphi(x)\,dx , \qquad (5.1.2)$$

where γ is a normalising constant determined by

$$\int_0^\infty \pi(t)\,dt = 1 \qquad (5.1.3)$$

and $h_e(\cdot)$ is the renewal density (Cox, 1962) of the e-events.

<u>Lemma 5.1.2</u> In Model 4.1, if

$$\pi_n(t,\tau) = \lim_{\Delta,\Delta',\Delta''\to 0} \text{Prob[the n-th i-event after the origin occurs in } (\tau,\tau+\Delta'), \text{ the first e-event after this n-th i-event occurs in } (t,t+\Delta'') \text{ and no r-event occurs in } (0,t) \mid \text{an r-event occurs in } (-\Delta,0)]/\Delta'\Delta'' , \quad (5.1.4)$$

then $\pi_n(t,\tau)$ is given by

$$\pi_1(t,\tau) = \pi(\tau)\,\varphi(t), \quad (5.1.5)$$

and, for $n \geqslant 2$,

$$\pi_n(t,\tau) = \int_0^\tau \pi_{n-1}(t,u)\,\psi(\tau-u)\,du$$

$$+ \int_0^\tau \varphi(t-v)\,dv \int_0^v \pi_{n-1}(v,u)\,\psi(\tau-u)\,du. \quad (5.1.6)$$

<u>Lemma 5.1.3</u> In Model 5.1, if

$$p_n(t) = \lim_{\Delta,\Delta'\to 0} \text{Prob[the first r-event after the origin occurs in } (t,t+\Delta') \text{ and n i-events occur in } (0,t) \mid \text{an r-event occurs in } (-\Delta,0)]/\Delta', \quad (5.1.7)$$

then $p_n(\cdot)$ is given by

$$p_0(t) = \varphi(t) \int_t^\infty \pi(u)\,du, \quad (5.1.8)$$

and, for $n \geqslant 1$,

$$p_n(t) = \int_0^t \varphi(t-u)\,du \int_0^u \pi_n(u,\tau)\,d\tau \int_{t-\tau}^\infty \psi(z)\,dz. \quad (5.1.9)$$

Based on these lemmas, the following theorem can be stated for Model 5.1:

<u>Theorem 5.1.1</u> In the basic deletion model (Model 5.1) the pdf of the interval between two successive uninhibited e-events

(r-events) is given by

$$f(t) = \sum_{n=0}^{\infty} p_n(t), \tag{5.1.10}$$

where $p_n(\cdot)$ is given by (5.1.8) and (5.1.9).
The special cases when $\varphi(\cdot)$ or $\psi(\cdot)$ is exponential are given by Theorems 5.1.2 and 5.1.3.

Theorem 5.1.2 In Model 5.1, when the i-events form a Poisson process with parameter μ, the Laplace transform of the pdf of the interval between two successive r-events is given by

$$f^*(s) = \frac{\varphi^*(s+\mu)}{1-\varphi^*(s) + \varphi^*(s+\mu)} . \tag{5.1.11}$$

Theorem 5.1.3 In Model 5.1, when the e-events form a Poisson process with parameter λ, the Laplace transform of the pdf of the interval between two successive r-events is given by

$$f^*(s) = \frac{\lambda}{s+\lambda} + \gamma\lambda\,[1-\psi^*(s+\lambda)]$$

$$[s\{-1+\psi^*(\lambda)\} - \lambda\{\psi^*(s+\lambda) - \psi^*(\lambda)\}]/$$

$$[(s+\lambda)^3\{1-\psi^*(s+\lambda) + \lambda\dot{\psi}^*(s+\lambda)\}], \tag{5.1.12}$$

where the dot denotes differentiation w.r.t. s.

Proof of Theorem 5.1.2 Since the i-events form a Poisson process with parameter μ,

$$\psi(t) = \mu e^{-\mu t}. \tag{5.1.13}$$

From (5.1.1),

$$\pi(t) = \mu e^{-\mu t}, \tag{5.1.14}$$

and, from (5.1.8),

$$p_0(t) = \varphi(t)\ e^{-\mu t}. \qquad (5.1.15)$$

Using (5.1.13) in (5.1.6)

$$\pi_n(z,w) = \int_0^w \pi_{n-1}(z,u)\ \mu e^{-\mu(w-u)} du +$$

$$\int_0^w \varphi(z-v)\ dv \int_0^v \pi_{n-1}(v,u)\ \mu e^{-\mu(w-u)} du. \qquad (5.1.16)$$

The following transform is introduced:

$$\bar{\pi}_n(z,\mu) = \int_0^z \pi_n(z,w)\ e^{\mu w} dw. \qquad (5.1.17)$$

Transforming (5.1.16),

$$\bar{\pi}_n(z,0) = \bar{\pi}_{n-1}(z,0) - e^{-\mu z}\ \bar{\pi}_{n-1}(z,\mu)$$

$$+ \int_0^z \varphi(z-v)\ e^{-\mu v}\ \bar{\pi}_{n-1}(v,\mu)\ dv$$

$$- e^{-\mu z} \int_0^z \varphi(z-v)\ \bar{\pi}_{n-1}(v,\mu)\ dv. \qquad (5.1.18)$$

Taking the Laplace transform of (5.1.18) w.r.t. z,

$$\bar{\pi}_n^*(s,0) = \bar{\pi}_{n-1}^*(s,0) - \pi_{n-1}^*(s+\mu,\mu)$$

$$[1-\varphi^*(s) + \varphi^*(s+\mu)]. \qquad (5.1.19)$$

Summing this over $n = 2$ to ∞,

$$\sum_{n=2}^{\infty} \bar{\pi}_n^*(s,0) = \sum_{n=2}^{\infty} \bar{\pi}_{n-1}^*(s,0)$$

$$- \sum_{n=2}^{\infty} \bar{\pi}_{n-1}^*(s+\mu,\mu)\ [1-\varphi^*(s) + \varphi^*(s+\mu)] \qquad (5.1.20)$$

which gives

$$\sum_{n=1}^{\infty} \bar{\pi}_n^*(s+\mu,\mu)=\frac{\bar{\pi}_1^*(s,0)}{1-\varphi^*(s)+\varphi^*(s+\mu)} . \qquad (5.1.21)$$

Now, $\bar{\pi}_1^*(s,0)$ can be obtained directly from (5.1.5) and (5.1.17) as

$$\bar{\pi}_1^*(s,0) = \varphi^*(s) - \varphi^*(s+\mu). \qquad (5.1.22)$$

From (5.1.9), (5.1.10), (5.1.13) and (5.1.14)

$$f(t) = \varphi(t)\, e^{-\mu t} + \sum_{n=1}^{\infty} \int_0^t \varphi(t-z)\, dz \int_0^z \pi_n(z,w)\, e^{-\mu(t-w)}\, dw \qquad (5.1.23)$$

and thence

$$f(t)=e^{-\mu t}[\varphi(t) + \int_0^t \varphi(t-z) \sum_{n=1}^{\infty} \bar{\pi}_n(z,\mu)\, dz]. \qquad (5.1.24)$$

This gives

$$f^*(s) = \varphi^*(s+\mu)\, [1 + \sum_{n=1}^{\infty} \bar{\pi}_n^*(s+\mu,\mu)] \qquad (5.1.25)$$

which, on using (5.1.21) and (5.1.22), leads to (5.1.11), a result first given by Ten Hoopen and Reuver (1965b) and subsequently obtained by many others (for example: Srinivasan and Rajamannar, 1970a;Lawrance, 1970).

This result can be derived more easily from fundamental considerations. Thus, let $h_r(\cdot)$ be the renewal density of the r-events (which form a renewal process because the i-events are Poisson). Then one can write

$$h_r(t) = \varphi(t)\, e^{-\mu t} + \int_0^t h_e(t-u)\, \varphi(u)\, e^{-\mu u}\, du \qquad (5.1.26)$$

from which

$$h_r^*(s) = \frac{\varphi^*(s+\mu)}{1-\varphi^*(s)} . \qquad (5.1.27)$$

Equation (5.1.11) is easily obtained from (5.1.26) since

$$f^*(s) = \frac{h_r^*(s)}{1+h_r^*(s)} . \qquad (5.1.28)$$

Proof of Theorem 5.1.3 Since the e-events are Poisson with parameter λ ,

$$\varphi(t) = \lambda e^{-\lambda t}, \qquad (5.1.29)$$

and the renewal density $h_e(\cdot)$ is given by

$$h_e(t) = \lambda . \qquad (5.1.30)$$

Equation (5.1.2) then reduces to

$$\pi(t) = \gamma \int_0^\infty \psi(t+w)\,(1-e^{-\lambda w})\, dw. \qquad (5.1.31)$$

Taking the Laplace transform w.r.t. t,

$$\pi^*(s) = \gamma \left[\frac{s-\lambda+\lambda\psi^*(s) - s\,\psi^*(\lambda)}{s(s-\lambda)} \right] \qquad (5.1.32)$$

from which γ is obtained as

$$\gamma = \frac{\lambda}{\lambda E_i - 1 + \psi^*(\lambda)} , \qquad (5.1.33)$$

using the fact that $\pi^*(0) = 1$ and taking the limit as $s \to 0$. Here E_i is the expected value of the interval between two successive i-events and is given by

$$E_i = \int_0^\infty t\,\psi(t)\, dt$$

$$= -\psi^{*\prime}(s) \Big|_{s=0} \qquad (5.1.34)$$

From (5.1.8),

$$p_0(t) = \lambda e^{-\lambda t} \int_t^{\infty} \pi(u)\, du, \tag{5.1.35}$$

so that

$$p_0^*(s) = \frac{\lambda[1-\pi^*(s+\lambda)]}{s+\lambda}. \tag{5.1.36}$$

Equation (5.1.5) becomes

$$\pi_1(t,\tau) = \lambda e^{-\lambda t} \pi(\tau) \tag{5.1.37}$$

and from (5.1.9),

$$p_1(t) = \int_0^t \lambda e^{-\lambda(t-z)} dz \int_0^z \pi_1(z,w) dw\, \chi(t-w) \tag{5.1.38}$$

where

$$\chi(t) = \int_t^{\infty} \psi(x)\, dx. \tag{5.1.39}$$

This can be simplified to

$$p_1(t) = \lambda^2 e^{-\lambda t} \int_0^t \pi(w)\chi(t-w)(t-w)\, dw \tag{5.1.40}$$

so that

$$p_1^*(s) = \lambda^2 \frac{\pi^*(s+\lambda)}{(s+\lambda)^2} [(s+\lambda)\dot{\psi}^*(s+\lambda) - \psi^*(s+\lambda) + 1]. \tag{5.1.41}$$

Using (5.1.29) in (5.1.6)

$$\pi_2(z,w) = \lambda e^{-\lambda z} [\int_0^w \pi(u)\psi(w-u)\, du$$

$$+ \lambda \int_0^w \pi(u)\psi(w-u)(w-u)\, du], \tag{5.1.42}$$

and from (5.1.9)

$$p_2(t) = \lambda^2 e^{-\lambda t} \int_0^t (t-w)\, dw \,[\int_0^w \pi(u)\, \psi(w-u)\, du$$

$$+ \lambda \int_0^w \pi(u)\, \psi(w-u)\, (w-u)\, du]\, \chi(t-w). \quad (5.1.43)$$

This yields

$$p_2^*(s) = \lambda^2 \frac{\pi^*(s+\lambda)}{(s+\lambda)^2} [\psi^*(s+\lambda) - \lambda\dot{\psi}^*(s+\lambda)]$$

$$[(s+\lambda)\dot{\psi}^*(s+\lambda) - \psi^*(s+\lambda) + 1]. \quad (5.1.44)$$

Similarly it can be shown that, for $n \geq 2$,

$$p_n^*(s) = \frac{\lambda^2 \pi^*(s+\lambda)}{(s+\lambda)^2} [(s+\lambda)\dot{\psi}^*(s+\lambda)$$

$$-\psi^*(s+\lambda) + 1][\psi^*(s+\lambda) - \lambda\dot{\psi}^*(s+\lambda)]^{n-1}. \quad (5.1.45)$$

Summing this over $n = 1$ to ∞,

$$\sum_{n=1}^{\infty} p_n^*(s) = \frac{\lambda^2 \pi^*(s+\lambda)}{(s+\lambda)^2} \frac{[(s+\lambda)\dot{\psi}^*(s+\lambda) - \psi^*(s+\lambda) + 1]}{[1-\psi^*(s+\lambda) + \lambda\dot{\psi}^*(s+\lambda)]}. \quad (5.1.46)$$

Adding this to (5.1.36) and using (5.1.32), and simplifying, results (5.1.12). This proves Theorem 5.1.3. The result obtained is the same as that due to Ten Hoopen and Reuver (1965b) except that γ here is equal to λ/c in their paper.

When neither $\varphi(\cdot)$ nor $\psi(\cdot)$ is exponential, $f^*(s)$ has an exceedingly complicated form. If one of the two input sequences has a gamma function for the interval pdf (the other having a general function), some simplification is possible (Pooi Ah Hin, 1974), though the expressions are still unwieldy. In any case, when $\psi(\cdot)$ is not exponential, the r-events do not form a renewal process and higher order properties have to be known. These are

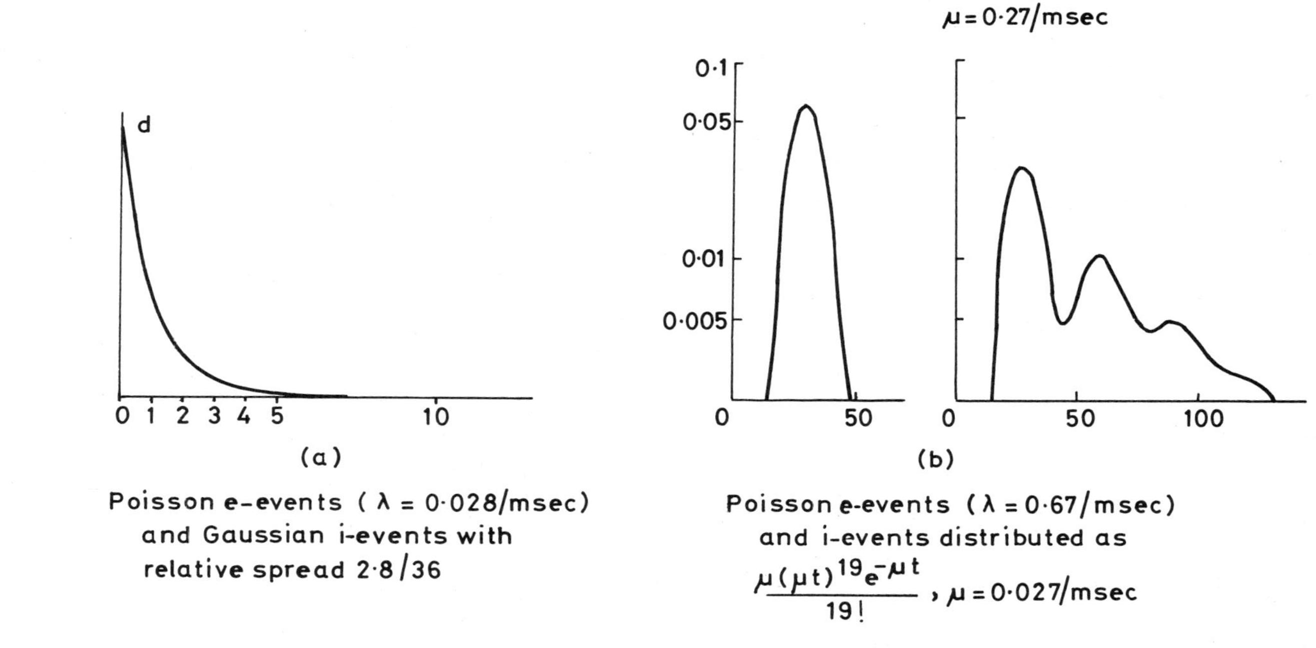

FIG. 5.1.1. INTERVAL DISTRIBUTIONS FROM MODEL 5-1 (Ten Hoopen and Reuver)

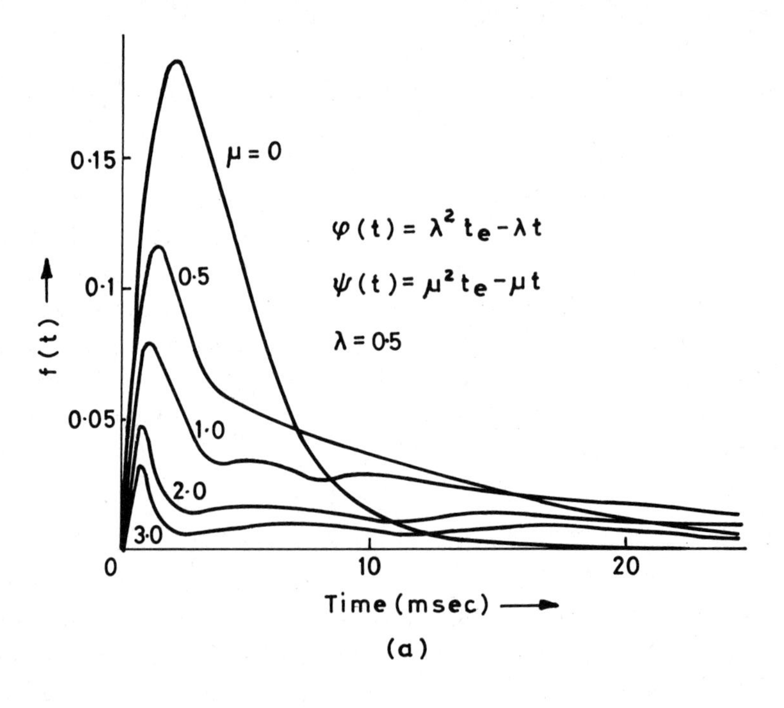

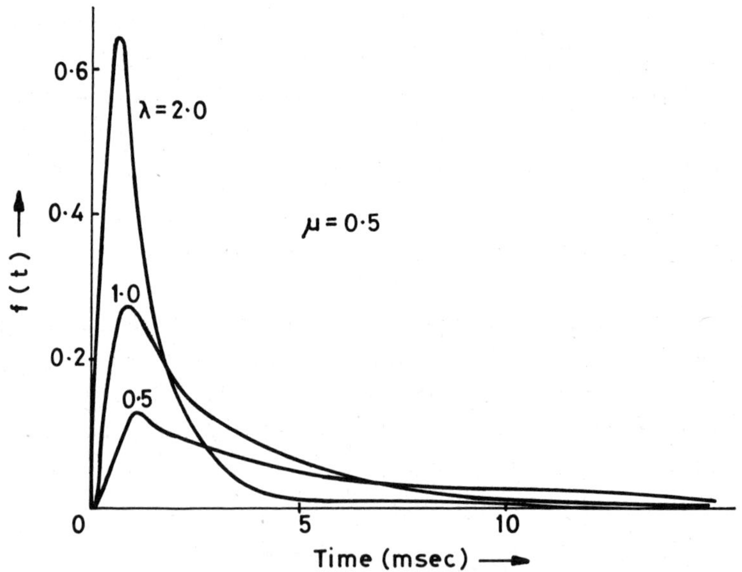

FIG. 5·1·2 INTERVAL DISTRIBUTIONS FROM MODEL 5·1 (Pooi Ah Hin, 1974).

discussed below.

The shapes of the pdf for the special cases discussed above are shown in Fig. 5.1.1. As observed by Bishop et al. (1964) in their simulation study, the interval pdf $f(\cdot)$ with $\psi(\cdot)$ exponential shows multimodal characteristics only when $\varphi(\cdot)$ is a high-order gamma function. When $\varphi(\cdot)$ is exponential, $f(\cdot)$ shows a bias towards the origin as it should. Ten Hoopen (1966a) has discussed both these cases in detail; the reader is referred to his paper. When neither $\varphi(\cdot)$ nor $\psi(\cdot)$ is exponential, multimodal characteristics can be obtained when $\varphi(\cdot)$ and $\psi(\cdot)$ are low-order gamma functions. This is seen in Fig. 5.1.2 which is taken from Pooi Ah Hin (1974).

5.1.2 Higher-order properties of the sequence of r-events (Srinivasan and Rajamannar, 1970a)

Considering the general case in which the e-events and the i-events form two independent stationary renewal point processes, the r-events form a non-Markovian non-renewal process. This can be studied through product densities. Here the first and second-order densities are derived. The higher-order densities can also be derived, though it is a tedious process. Similar studies relating to the interaction of stationary processes have been done by Lawrance (1970).

Lemma 5.1.4 In Model 5.1, if

$F_1(\cdot)$ is the product density of degree one of the i-events occurring at u, conditional on an r-event occurring at the origin, $\pi(\cdot)$ is given by (5.1.2) and (5.1.3), and $h_i(\cdot)$ the renewal density of the i-events, then the product density $F_1(\cdot)$ satisfies

$$F_1(u) = \pi(u) + \int_0^u \pi(u')\, h_i(u-u')\, du'. \qquad (5.1.47)$$

One can then prove

Theorem 5.1.4 The first-order product density $h_1(\cdot)$ of the r-events in the basic deletion model (Model 5.1) is given by

$$h_1(t) = h_e(t) \int_t^{\infty} \pi(t')\, dt' \; +$$

$$\int_0^t F_1(u)\, du \int_u^t h_e(v)\, \varphi(t-v)\, \chi(t-u)\, dv, \qquad (5.1.48)$$

where $h_e(\cdot)$ is the renewal density (product density of degree one) of the e-events and $\chi(\cdot)$ is given by (5.1.39).

Proof It is sufficient to observe that the set of events leading to an r-event in $(t, t+dt)$ is the union of two mutually exclusive sets corresponding to a response occurring in $(t, t+dt)$ with and without an i-event occurring in $(0,t)$. Equation (5.1.48) is accordingly written, using Lemma 5.1.4.

Thus an explicit expression for the first-order product density is now available. It may be observed that in deriving this there is no need to know $f(\cdot)$, the interval pdf. On taking the limit as $t \to \infty$ in equation (5.1.48), it is possible to prove

$$\lim_{t \to \infty} h_1(t) = \int_0^{\infty} \chi(u)\, du \int_0^u \varphi(v)\, dv / E_i E_e, \qquad (5.1.49)$$

where E_i is given by equation (5.1.34) and E_e is the expected value of the interval between two successive e-events and is given by a similar expression. When the e-events are Poisson, i.e., $\varphi(\cdot)$ is given by (5.1.29),

$$\lim_{t \to \infty} h_1(t) = \lambda + \frac{1-\psi^*(\lambda)}{E_i}. \qquad (5.1.50)$$

Theorem 5.1.5 In the basic deletion model (Model 5.1), the product density of degree two $h_2(\cdot,\cdot)$ of the r-events is given by

$$h_2(x,y) = \int_y^{\infty} \pi(u)\, h_e(x,y)\, du$$

$$+ \int_0^x F_1(u)\, du\, \chi(y-u) \int_u^x h_e(v)\, \varphi(x-v) h_e(y-x)\, dv$$

$$+ \int_x^y \pi(u)\, \chi\, (y-u)\, du \int_u^y h_e\, (x,v)\, \varphi(y-v)\, dv$$

$$+ \int_x^y \pi(u)\, du \int_u^y h_i(w-u)\, \chi\, (y-w)\, dw \int_w^y h_e(x,z)\, \varphi(y-z)\, dz$$

$$+ \int_0^x F_1(u)\, du \int_x^y \psi\, (w-u)\, \chi\, (y-w)\, dw \int_w^y dy' \int_u^x h_e(x')$$

$$\varphi(x-x')\, h_e\, (y'-x)\, \varphi\, (y-y')\, dx'$$

$$+ \int_0^x F_1(u)\, du \int_x^y \psi\, (w-u)\, dw \int_w^y h_i(v-w)\, \chi\, (y-v)\, dv \int_v^y dy'$$

$$\int_u^x h_e(x')\, \varphi(x-x')\, h_e(y')\, \varphi(y-y')\, dx', \qquad (5.1.51)$$

where $h_{e_2}\ (\cdot,\cdot)$ is the product density of degree two of the e-events.

Proof It is first observed that the set of events leading to a response occurring in $(x,x+dx)$ and a response in $(y,y+dy)$, $y > x$, is the union of the following mutually exclusive classes:

i) the set of events in which there is no i-event in $(0,y)$,

ii) the set in which an i-event occurs in $(0,x)$ and none in (x,y),

iii) the set in which no i-event occurs in $(0,x)$ but an i-event occurs in (x,y), and

iv) the set in which an i-event occurs in both $(0,x)$ and (x,y).

Equation (5.1.51) is written taking these four possibilities into account.

Since the process is stationary, the stationary second-order product density $\rho(\cdot)$ is obtained thus:

$$\rho(t) = \lim_{\substack{t_1, t_2 \to \infty \\ t_2 - t_1 \to t}} h_2(t_1, t_2). \tag{5.1.52}$$

The spectral response can then be obtained from (5.1.52).

<u>Lemma 5.1.5</u> In Model 5.1, if the e-events form a Poisson process with parameter λ, and

$$Q(t) = \lim_{\Delta, \Delta' \to 0} \text{Prob [an i-event occurs in } (t, t+\Delta') \text{ and no r-event occurs in } (0,t) \mid \text{an i-event occurs in } (-\Delta, 0)]/\Delta',$$

then $Q(\cdot)$ satisfies the equation

$$Q(t) = \psi(t)\,[e^{-\lambda t} + \lambda t e^{-\lambda t}] + \int_0^t Q(t')\,\psi(t-t')\,[e^{-\lambda(t-t')} + \lambda(t-t')\,e^{-\lambda(t-t')}]\,dt'. \tag{5.1.53}$$

<u>Lemma 5.1.6</u> In Model 5.1, if the e-events form a Poisson process with parameter λ, and

$$q(t) = \lim_{\Delta, \Delta' \to 0} \text{Prob [an i-event occurs in } (t, t+\Delta') \text{ and no r-event in } (0,t) \mid \text{an r-event occurs in } (-\Delta, 0)]/\Delta',$$

then $q(\cdot)$ satisfies the equation

$$q(t) = \pi(t)\,e^{-\lambda t} + \int_0^t q(t')\,\psi(t-t')[1+\lambda(t-t')]\,e^{-\lambda(t-t')}\,dt'. \tag{5.1.54}$$

<u>Theorem 5.1.6</u> In Model 5.1, when the e-events form a Poisson process with parameter λ, the bivariate pdf $f(\cdot,\cdot)$ of two successive intervals separated by r-events is given by

$$f(t_1, t_2) = \lambda^2 e^{-\lambda t_2} \int_{t_2}^{\infty} \pi(t')\,dt'$$

$$+\lambda e^{-\lambda t_2} \int_{t_1}^{t_2} \pi(t')\, \lambda^2(t_2-t')\, \chi\,(t_2-t')\, dt'$$

$$+\lambda\; e^{-\lambda t_1} \int_{t_1}^{t_2} \pi(t')\, e^{-\lambda(t'-t_1)}\, dt' \int_{t'}^{t_2} Q(u-t')$$

$$e^{-\lambda(t_2-u)}\, \lambda^2(t_2-u)\, \chi\,(t_2-u)\, du$$

$$+\int_0^{t_1} q(u)\, \lambda^2(t_1-u)\, \chi\,(t_2-u)\, \lambda\, e^{-\lambda(t_2-u)}\, du$$

$$+\int_0^{t_1} q(u)\, e^{-\lambda(t_2-u)}\, \lambda^2(t_1-u)\, du$$

$$\int_{t_1}^{t_2} \psi\,(v-u)\, \lambda^2(t_2-v)\, \chi\,(t_2-v)\, dv$$

$$+\int_0^{t_1} q(u)\; e^{-\lambda(t_1-u)}\, \lambda^2(t_1-u)\, du$$

$$\int_{t_1}^{t_2} \psi\,(v-u)\; e^{-\lambda(v-t_1)}\, dv \int_v^{t_2} Q(w-v)\, e^{-\lambda(t_2-w)}$$

$$\lambda^2(t_2-w)\, \chi\,(t_2-w)\, dw, \qquad (5.1.55)$$

where $Q(\cdot)$ and $q(\cdot)$ are given by (5.1.53) and (5.1.54).

<u>Proof of Theorem 5.1.6</u> As before, the theorem is proved by classifying the r-events occurring at t_1 and t_2 into 4 mutually exclusive classes:

i) the r-events occurring at t_1 and t_2 are the first and and second e-events following an r-event at the origin

ii) the r-event at t_1 is the first e-event after the origin and at t_2 is the second or subsequent e-event after t_1,

iii) the r-event at t_1 is a second or subsequent e-event after $t=0$, and the r-event at t_2 is the first one after t_1,

iv) the r-event at t_1 is a second or subsequent e-event after $t = 0$, and the r-event at t_2 is a second or subsequent e-event after t_1.

The terms on the right hand side of equation (5.1.55) correspond to these four possibilities.

5.1.3 Extended versions of Model 5.1

In the basic deletion model it is assumed that an i-event has its inhibitory effect only on the first e-event following it. Inhibition is, however, due to the chemical nature of the transmitter present, and the chemical reactions in the synaptic region are reversible. This leads to many possibilities:

1) an i-event has its effect lasting for a certain lifetime which is certainly not infinity,

2) this effect will be felt on all e-events arriving within that lifetime.

Accordingly models have been proposed by Pooi Ah Hin (1974) and Rade (1972) who have postulated multiple pre-inhibition of more than one e-event by an i-event preceding. Considering the mortal nature of pre-inhibition, Coleman and Gastwirth (1969) and Srinivasan and Rajamannar (1970c) have assumed that the inhibitory effect lasts for a length of time T during which e-events may be deleted in prescribed ways. T could be a constant or a r.v. In this section, only the former class is considered and in this Rade's model alone is discussed. Since the concept of lifetime is similar to that of dead time in a counter the latter class of models is studied in Chapter 7.

Model 5.2 (Rade, 1972)

This model has the following features:

1) The e-events form a renewal process with interval pdf $\varphi(\cdot)$.

2) The i-events are Poisson with parameter μ.

3) One or more i-events delete the next k e-events with probability p_k, $\sum_{k=0}^{\infty} p_k = 1$. An i-event arriving during a time period when e-events are deleted by previous i-events has no effect.

4) Undeleted e-events (r-events) form the output in the model.

Since the i-events are Poisson, the r-events form a renewal process, and hence the r-process is completely described by the interval pdf $f(\cdot)$. The following theorem which gives an expression for the Laplace transform of $f(\cdot)$ is stated without proof.

Theorem 5.1.7 The Laplace transform of the interval pdf of the r-process in Model 5.2 is given by

$$f^*(s) = \frac{p_0 \, \varphi^*(s) + (1-p_0)\, \varphi^*(s+\mu)}{1 - [\varphi^*(s) - \varphi^*(s+\mu)] \sum_{k=1}^{\infty} p_k \, [\varphi^*(s)]^{k-1}} \, . \tag{5.1.56}$$

Corollary When $k = 1$, Model 5.2 reduces to the basic deletion model (Model 5.1).

This follows from the fact that Model 5.1 is obtained from Model 5.2 by observing that

$$\begin{aligned} p_k &= 1, \qquad k = 1 \\ &= 0, \qquad k \quad 0 \end{aligned} \tag{5.1.57}$$

5.2 Models with dependent interaction of excitatory and inhibitory sequences - Models 5.3 and 5.4

In Section 5.1 it has been assumed that the e- and i-sequences are independent of each other. This is only an approximation, since neurons are interconnected and the firing sequences are dependent. In certain parts of the nervous system, this dependence is extremely pronounced. For example, in the spinal cord, inhibition is known to occur by local feedback through interneurons called Renshaw cells. The latter, by the hyperpolarisation they cause, inhibit the surrounding motoneurons (Ochs, 1965, p. 333 and 356). This effect is somewhat similar to lateral inhibition in which a sensory neuron inhibits its neighbour by an amount related to the distance between the two. Such a mechanism is responsible for contrast in vision (Ratliff, 1961).

In an attempt to model such neural phenomena, Ten Hoopen and Reuver (1968) proposed a scheme of dependent interaction of e- and i-events in which one of the streams is controlled by the other. There are two models (Models 5.3 and 5.4) and these are discussed below.

Models 5.3 and 5.4 (Ten Hoopen and Reuver, 1968; Srinivasan and Rajamannar, 1970b)

The hypothetical neuron has two kinds of input sequences - excitatory (e-events) and inhibitory (i-events) - incident on it and has the following properties:

i) The e-sequence is a renewal process with interval pdf $\varphi(\cdot)$.

ii) An e-event starts a sequence of i-events, also forming a renewal process with pdf $\psi(\cdot)$.

iii) The i-sequence is independent of the e-process, except for the trigger control exercised by the triggering e-event, i.e., the time of occurrence of the latter is the origin of the i-sequence started.

iv) Every i-event pre-inhibits the first e-event that follows it.

v) In Model 5.3 every e-event, pre-inhibited or not, starts an i-sequence while simultaneously stopping the i-sequence started by the previous e-event.

vi) In Model 5.4, only uninhibited e-events can start an i-sequence while simultaneously stopping the i-sequence started by the previous e-event.

vii) In both the models, the sequence of uninhibited e-events (r-events) is the neuron output spike train.

It is observed that the r-events form a renewal process since the origin of an i-sequence is always an e-event, and an r-event coincides with an e-event. This means that the r-event sequence is only the e-sequence with some of the e-events 'filtered out'. The interval pdf of the r-events therefore completely characterises the r-sequence.

<u>Theorem 5.2.1</u> In Model 5.3, the renewal density $h_r(\cdot)$ of the uninhibited e-events (r-events) is given by

$$h_r(t) = \varphi(t) \int_t^{\infty} \Psi(t)\, dt + \int_0^t h_e(\tau)\, \varphi(t-\tau) d\tau \int_{t-\tau}^{\infty} \psi(u) du, \qquad (5.2.1)$$

where $h_e(\cdot)$ is the renewal density of the e-events.

The proof of the theorem is simple and omitted.

<u>Theorem 5.2.2</u> In Model 5.3, the Laplace transform of the interval pdf of the r-events $f(\cdot)$ is given by

$$f^*(s) = \frac{K^*(s)}{1 + K^*(s) - \varphi^*(s)} \qquad (5.2.2)$$

where

$$K(t) = \varphi(t) \int_t^{\infty} \Psi(u)\, du. \qquad (5.2.3)$$

<u>Proof</u>: From the renewal equation (Cox, 1962),

$$f^*(s) = \frac{h_r^*(s)}{1 + h_r^*(s)}. \qquad (5.2.4)$$

From Theorem 5.2.1,

$$h_r^*(s) = K^*(s)\, [1 + h_e^*(s)], \qquad (5.2.5)$$

and $h_e^*(s)$ is obtained from the renewal equation as

$$h_e^*(s) = \frac{\varphi^*(s)}{1 - \varphi^*(s)} . \tag{5.2.6}$$

Combining the last three equations, equation (5.2.2) is obtained.

Corollary: When the e-events are Poisson with parameter λ,

$$f^*(s) = \frac{\lambda[1 - \psi^*(s+\lambda)]}{s + \lambda - \lambda\psi^*(s+\lambda)} . \tag{5.2.7}$$

Theorem 5.2.3 In Model 5.4, if the e-events form a Poisson process with parameter λ the Laplace transform of the interval pdf $f(\cdot)$ of the uninhibited e-events (r-events) is given by

$$f^*(s) = \frac{\lambda}{\lambda + s} + \frac{\lambda s}{(\lambda + s)^2} \; \frac{\psi^*(s+\lambda)\,[\psi^*(s+\lambda)-1]}{1 - \psi^*(s+\lambda) + \lambda\dot{\psi}^*(s+\lambda)} . \tag{5.2.8}$$

Proof: First it is observed that Model 5.4 is not very much different from Model 5.1. The only difference here is that the origin of i-sequence is an r-event, whereas in the independent interaction model, the i-events form a delayed renewal process with respect to an r-event. It is then easy to see that the analysis for Model 5.4 is similar with the difference that equation (5.1.2) is simply

$$\pi(t) = \psi(t), \tag{5.2.9}$$

and this is to be used in the analysis that follows equation (5.1.2). Thus when $\varphi(\cdot)$ is exponential, adding equations (5.1.36) and (5.1.46) and using equation (5.2.9), equation (5.2.8) is obtained.

In passing, it may be mentioned here that the result given by Srinivasan and Rajamannar (1970b) is different because they have assumed that every e-event, pre-inhibited or not, stops the previous sequence. In this case the expression for the pdf is much simpler. The reason is that two successive r-events can be interrupted by not more than one e-event. This leads to

<u>Theorem 5.2.4</u> In Model 5.4, if it is assumed that every e-event, pre-inhibited or not, stops the previous i-sequence, the interval pdf $f(\cdot)$ of the r-events is given by

$$f(t) = K(t) + \int_0^t \varphi(\tau)\,\varphi(t-\tau)\,d\tau \int_0^\tau \psi(u)du \quad (5.2.9)$$

where $K(\cdot)$ is given by equation (5.2.3).

Several modifications of Models 5.3 and 5.4 including the concept of 'lifetime' of inhibitories have been made; these versions have been studied using the methods of counter theory (Srinivasan and Rajamannar, 1970c) and are discussed in Section 7.3.

5.3 Discussion

The deletion model has been criticised for ignoring the time evolution of the membrane potential. Nevertheless, the model should not be totally rejected, because pre-synaptic inhibition does occur in the sensory system and sometimes in the higher nerve centers of lower species as well (Eccles, 1973).

The drawback relating to the integrating effect over successive post-synaptic potentials can be removed if it is recognised that the post-synaptic effect of an r-event is to be considered an EPSP and not a spike discharge. The post-synaptic membrane potential then evolves with successive r-events. A very simple integrator model would consist of a threshold firing scheme in which the neuron stores the EPSPs due to r-events and fires whenever the number in store reaches N, where N is the threshold number. The membrane potential immediately thereafter drops back to zero. Thus the interval pdf $P(\cdot)$ of the output spike sequence of such an integrator neuron is given by

$$P(t) = f_N(t). \quad (5.3.1)$$

Lee (1974) has studied such integrator models starting with the deletion model extensions due to Coleman and Gastwirth (1969). A more realistic scheme takes into account the random nature of the PSP, i.e., the neuron does not fire exactly after N r-events, because the post-synaptic change due to an r-event is itself a random variable (see p. 11). A model including this property along with a parameter representing natural decay of the membrane potential to the rest level in the absence of r-events is discussed in Section 9.2.2. Similar modifications of the dependent interaction models (Model 5.3 and Model 5.4) have been considered and are discussed in Section 9.4.

CHAPTER 6

DIFFUSION MODELS

When a neuron has a large number of synapses, the sequence of inputs can be considered to approximate to a Poisson process. This effect of superposition has been discussed at length in Chapter 4. In addition, if the change in the membrane potential due to an input impulse is small compared to the difference between the threshold and the resting level, the time course of the membrane potential can be described by a Fokker-Planck type of equation. Such processes in which the arrivals of impulses are random and the change in the state of the stochastic process due to these impulses is small are called _diffusion processes_. The terminology arises from the analogy with diffusion phenomena in physics and chemistry. The equation describing such processes is known as the _diffusion equation_ (and also as the _heat equation_ in thermodynamics). Due to its wide application there exists a large body of literature, which can be used advantageously in modelling.

The study of the diffusion process in the field of probability theory has its origin in the classical models of Brownian motion in which particles move randomly in small rapid steps in three dimensions. Many biological phenomena are described by equations similar to those that describe Brownian motion and these have the common property that the changes of state of some biological variable (that has the properties of a random process) are small and take place at completely random instants of time. The words 'completely random' may be translated into the language of random processes as 'mutually independent'. Or, viewed at another angle, if the changes are instantaneous, the sequence of 'completely random' instants of time at which they take place is a Poisson process. In modelling a neuron these two properties have a corresponding physical significance (mentioned at the beginning of this chapter). The assumption of these properties is really an approximation as far as neurons are concerned. This is referred to as the diffusion approximation. Under this approximation, several models of neurons have been proposed in the literature.

In the next section, the mathematical theory of diffusion processes is presented, first as the limit of a random walk and then in more general terms, the treatment following Lindley (1965) and Gnedenko (1969). Against this background the different diffusion models of neuron firing are described in Section 6.2. The chapter concludes with an appraisal of these models.

6.1 The diffusion equation

6.1.1 The diffusion process as the limit of a random walk

Consider a particle starting at n=0 and executing a random walk on the real integers without any boundaries. Let the probability of a step in the positive direction be p and that in the negative $q = 1-p$. Then if X_n is the random variable corresponding to the position of the particle after the n-th step,

$$X_n = X_{n-1} + Z, \quad n \geqslant 1, \tag{6.1.1}$$

where Z is the random variable corresponding to the step at any position with the discrete distribution

$$\text{prob } (Z = 1) = p$$

and

$$\text{prob } (Z = -1) = q \tag{6.1.2}$$

and the initial state is described by

$$\text{prob } (X_0 = 0) = 1. \tag{6.1.3}$$

Thus X_n is the sum of n iid r.v.s. Z with mean

$$E(Z) = p-q \tag{6.1.4}$$

and variance

$$\text{Var}(Z) = 4pq. \tag{6.1.5}$$

From the above description it is clear that the random walk $\{X_n\}$ is a simple Markov chain. If

$$\pi_n(r) = \text{prob}\left\{ X_n = r \right\}, \tag{6.1.6}$$

then

$$\pi_{n+1}(r) = p\,\pi_n(r-1) + q\,\pi_n(r+1) \tag{6.1.7}$$

with the initial condition

$$\pi_0(0) = 1. \tag{6.1.8}$$

This set of difference equations can be solved to obtain an expression for $\pi_n(r)$. However the present aim is not to solve for $\pi_n(r)$ but to find the limiting behaviour of $\pi_n(r)$ by modifying equation (6.1.7) to bring in a time parameter. Before doing this a more instructive way of arriving at the limiting distribution of $\pi_n(r)$ is considered below.

The random walk described above is equivalent to a sequence of tosses of a coin with the probability of a head turning up being p and of a tail q. If j heads show up in n trials then the number of tails must be n-j. Therefore the net number of heads r is given by

$$\begin{aligned} r &= j - (n-j) \\ &= 2j - n, \end{aligned} \tag{6.1.9}$$

from which

$$j = \frac{r+n}{2}, \tag{6.1.10}$$

where j can be a positive integer only. Now, the sequence of tosses is a Bernoulli sequence. Hence $\pi_n(r)$ has a binomial distribution. It is easy then to arrive at

Theorem 6.1.1 In a simple random walk on the real integers starting from zero with the probability of a step to the right being p and a step to the left being $q = 1-p$, the probability that the particle is in position r in n steps is given by

$$\pi_n(r) = \binom{n}{\frac{n+r}{2}} p^{\frac{n+r}{2}} q^{\frac{n-r}{2}} . \qquad (6.1.11)$$

At this point, the form of the limiting distribution of $\pi_n(r)$ as $n \to \infty$ can be shown to be normal. This property is very useful because of the ease with which $\pi_n(r)$ can be obtained for large n from tabulated values of the normal distribution. This limiting property is given formally by

Theorem 6.1.2 (DeMoivre-Laplace theorem).
As $n \to \infty$, the binomial distribution

$$\pi_n(r) = \binom{n}{\frac{n+r}{2}} p^{\frac{n+r}{2}} q^{\frac{n-r}{2}}$$

tends to

$$\frac{1}{\sqrt{2\pi npq}} e^{-\frac{x^2}{2}} \qquad (6.1.12)$$

uniformly in all r for which x lies in some finite interval. The proof is somewhat lengthy and is omitted. See Feller (1960, pp 168-173).

The limiting behaviour of $\pi_n(r)$ as given by the difference equation (6.1.7) is now considered. The aim is to obtain this as a differential equation by using appropriate limit arguments. The solution of this equation is then shown to be a normal distribution (cf. equation 6.1.12).

Consider each step to be of length Δx (rather than unity) and the time taken between two successive steps Δt. Then the net displacement of the particle is $\frac{x}{\Delta x}$ while the number of steps taken is $\frac{t}{\Delta t}$. Let the random variables X_n and Z with the introduction of the quantities t and x be represented by $\underline{X}_n$ and $\underline{Z}$. Since $\underline{X}_n$ is the sum of n iid r.v.s $\underline{Z}$, the mean and the variance of $\underline{X}_n$ are given by

<u>Theorem 6.1.3</u> In the simple random walk described above,

$$E(\underline{X}_n) \cong (p-q)\ \Delta x\ \frac{t}{\Delta t} \tag{6.1.13}$$

and

$$\operatorname{Var}(\underline{X}_n) \cong 4pq\ (\Delta x)^2\ \frac{t}{\Delta t}\,. \tag{6.1.14}$$

The proof is simple because the mean and the variance of the sum of independent r.v.s are the sum of the means and the sum of the variances of these r.v.s respectively.

Now let Δx and Δt tend to zero. For a meaningful limit, the mean and the variance must be finite, with the variance not vanishing. Therefore let

$$4pq\ \frac{(\Delta x)^2}{\Delta t} \to \sigma^2 > 0, \tag{6.1.15}$$

and subtract from each step $\pm\Delta x$ the quantity $(p-q)\ \Delta x$ so that the mean of the distribution of $\underline{X}_n$ is zero. With this transformation the random variable describing the random walk becomes

$$\underline{X}_n' = \underline{X}_n - (p-q)\ \Delta x\ \frac{t}{\Delta t}\,, \tag{6.1.16}$$

and the difference equation (6.1.7) is rewritten as

$$f(x,t+\Delta t) = p\ f(x-2q\,\Delta x,\ t) + q\ f(x+2p\,\Delta x, t). \tag{6.1.17}$$

Subtracting $f(x,t)$ from both sides, dividing by Δt and taking the limit as $\Delta t \to 0$ and $\Delta x \to 0$ according to (6.1.15), one gets

$$f_t(x,t) = \frac{1}{2}\,\sigma^2\, f_{xx}(x,t) \tag{6.1.18}$$

which is the Fokker-Planck equation or the diffusion equation. The quantity $\frac{1}{2}\sigma^2$ is called the <u>diffusion coefficient</u>. The differential equation (6.1.18) with the initial condition

$$f(x,0) = \delta(x) \tag{6.1.19}$$

and with no boundaries is an initial value problem, the solution of which is well-known (see, for example, Sneddon (1957)) and is given by

<u>Theorem 6.1.3</u> The diffusion equation (6.1.18) with the condition (6.1.19) has the solution

$$f(x,t) = \frac{1}{\sqrt{2\pi t\sigma^2}} \exp\left(-\frac{x^2}{2\sigma^2 t}\right). \tag{6.1.20}$$

This is of the same form as (6.1.12).

In the above derivation, the mean was made zero by subtracting $(p-q)\,\Delta x$ from each step $\pm\,\Delta x$. If instead the mean is kept finite, and non-zero by letting

$$(p-q)\,\Delta x\,\frac{t}{\Delta t} \rightarrow ct \tag{6.1.21}$$

(in effect, $p-q$ is $o(\Delta x)$, considering (6.1.15)) then the diffusion equation takes the form

$$f_t(x,t) = -c\,f_x(x,t) + \frac{1}{2}\sigma^2 f_{xx}(x,t). \tag{6.1.22}$$

This is the <u>one dimensional diffusion equation with drift</u>, where c is called the <u>drift parameter</u>. The solution of this equation is obtained by a suitable transformation.

<u>Theorem 6.1.4</u> The solution of the one-dimensional diffusion equation with drift (6.1.22) with $f(x,0)=\delta(x)$ is given by

$$f(x,t) = \frac{1}{\sqrt{(2\pi\sigma^2 t)}} \exp\left(-\frac{(x-ct)^2}{2\sigma^2 t}\right). \tag{6.1.23}$$

<u>Proof</u>: As mentioned before, the theorem is proved by using a transformation, viz.,

$$f(x,t) = Q(x,t)\exp\left(\frac{cx}{\sigma^2} - \frac{c^2 t}{2\sigma^2}\right), \tag{6.1.24}$$

which on substitution in (6.1.22) leads to

$$Q_t = \frac{1}{2}\sigma^2 Q_{xx} \qquad (6.1.25)$$

with the initial condition

$$Q(x,0) = \delta(x). \qquad (6.1.26)$$

The solution to this initial value problem is given by Theorem 6.1.3. On making use of (6.1.24), equation (6.1.23) follows.

Before passing on to the study of the diffusion process in more general terms it is pertinent to remark here that the normal distribution form of the solution obtained above is to be expected as a result of the Central Limit Theorem because X_n is the sum of independent r.v.s with finite mean and variance.

6.1.2 General theory of diffusion processes

Consider a stochastic process $X(t)$ with the following properties:

1) $X(t)$ is a stationary Markov process.

2) If $F(x,t|x_0,t_0) = \text{Prob}\ [X(t) \leqslant x | X(t_0) = x_0,\ t_0 < t]$, (6.1.27)

then

$$F_{x_0}(x,t|x_0,t_0) \text{ and } F_{x_0 x_0}(x,t|x_0,t_0)$$

exist and are continuous for all x,t,x_0 and $t_0 < t$.

3) $X(t)$ is continuous, i.e., for an arbitrary $\Delta_1 > 0$

$$\lim_{\Delta t \to 0} \frac{1}{\Delta t} \int_{|x-x_0| \geqslant \Delta_1} d_x F(x,t+\Delta t | x_0,t) = 0 . \qquad (6.1.28)$$

4) For an arbitrary $\Delta_2 > 0$,

$$\lim_{\Delta t \to 0} \frac{1}{\Delta t} \int_{|x-x_0| < \Delta_2} (x-x_0)\, d_x F(x,t+\Delta t | x_0,t)$$

$$= a(x,t) \qquad (6.1.29)$$

and

$$\lim_{\Delta t \to 0} \frac{1}{\Delta t} \int_{|x-x_0|<\Delta_2} (x-x_0)^2 \, d_x F(x,t+\Delta t \mid x_0,t)$$

$$= b(x,t) \qquad (6.1.30)$$

both exist, the limits being uniform in x.

5) There exists a pdf

$$f(x,t \mid x_0,t_0) = F_x(x,t \mid x_0,t_0). \qquad (6.1.31)$$

6) The derivatives

$$f_t(x,t \mid x_0,t_0), \quad \frac{\partial}{\partial x}[a(x,t)\, f(x,t \mid x_0, t_0)]$$

and

$$\frac{\partial^2}{\partial x^2}[b(x,t)\, f(x,t \mid x_0, t_0)]$$

exist and are continuous.

Assumptions (1) and (3) are a statement of the diffusion approximation. Compare (4) with (6.1.15) and (6.1.21); it states that the mean and the variance of the changes of state are finite though they may be functions of both t and the state of X(t). Conditions (2), (5) and (6) ensure the formation of the differential equations given below.

Theorem 6.1.5 If conditions (1) - (4) stated above are satisfied by X(t), then $F(x,t \mid x_0,t_0)$ satisfies

$$F_{t_0}(x,t \mid x_0,t_0) = -\, a(x_0,t_0)\; F_{x_0}(x,t \mid x_0,t_0)$$

$$-\, \tfrac{1}{2}\, b(x_0,t_0)\; F_{x_0x_0}(x,t \mid x_0, t_0), \qquad (6.1.32)$$

and if all conditions except (2) are satisfied then the pdf $f(x,t \mid x_0,t_0)$ satisfies

$$f_t(x,t|x_0,t_0) = -\frac{\partial}{\partial x}[a(x,t)\, f(x,t|x_0,t_0)]$$

$$+\frac{1}{2}\frac{\partial^2}{\partial x^2}[b(x,t)\, f(x,t|x_0,t_0)]. \qquad (6.1.33)$$

Equations (6.1.32) and (6.1.33) are respectively called the <u>forward</u> and <u>backward Kolmogorov equations</u>. For the proof, see Gnedenko (1969, p.310). Here it is useful to observe that in (6.1.32) $F(x,t|x_0,t_0)$ can be replaced by $f(x,t|x_0,t_0)$ by differentiating $F(x,t|x_0,t_0)$ w.r.t. x. Also, in (6.1.33) if

$$a(x,t) = c \qquad (6.1.34)$$

and

$$b(x,t) = \frac{1}{2}\sigma^2 \qquad (6.1.35)$$

then (6.1.22) is obtained.

Theorem 6.1.5 gives the general diffusion equation for a process for which the diffusion approximation holds. In the next section some diffusion models that have been proposed for the single neuron are studied. Mathematically this involves solution of (6.1.32) and (6.1.33) for specific forms of a (x,t) and b(x,t) and under prescribed initial and boundary conditions.

6.2 Diffusion models for neuron firing sequences

As has been pointed out earlier in this chapter, under the diffusion approximation the evolution of the neuron membrane potential with time can be described by a diffusion equation. The neuron fires when the membrane potential crosses the threshold level after which it is reset to zero and repeats the process; therefore the diffusion process for a neuron becomes a first-passage-time problem that is a boundary value problem with the threshold level as the boundary. Hence a diffusion model reduces to the solution of a diffusion equation with prescribed boundary conditions. There have been several formulations, notably by Gluss (1967), Stein (1967), Roy and Smith (1969), Griffith (1971), Capocelli and Ricciardi (1973), and Clay and Goel (1973). The basis of all these models is the diffusion approximation, and the differences among them lie either in the approach to the problem or in the assumption of additional features representing neurophysiological properties. The precursor of these models was the one due to Gerstein and Mandelbrot (1964) who used a limiting form of the random walk model (which has been discussed in Section 6.1.1) and solved the corresponding boundary value problem. Later, the decay of the membrane potential in the absence of input impulses was included in the diffusion equation by Gluss (1967) who obtained the Laplace transform solution of the first passage time problem. Roy and Smith (1969) followed this up and obtained an expression for the mean of the first passage time. Griffith (1971) derived the diffusion equation with exponential decay in the absence of inputs from fundamental considerations and extended the model further. The diffusion approximation was examined by Capocelli and Ricciardi (1973) who also considered the case of finite width input impulses. In an attempt to include refractoriness in the diffusion model, Clay and Goel (1973) considered the return of the threshold to the normal level after a firing and solved special cases of the diffusion equation with specific forms of time varying threshold. Some of these models are discussed in this section. Since all of them have some common features, to avoid repetition the basic diffusion model is assumed to have the following properties (unless otherwise stated):

a) The membrane potential $X(t)$ is set to a level x_0 immediately after a firing at t_0 (usually assumed to be zero).

b) The neuron receives excitatory impulses (e-events) in a Poisson stream with rate λ and inhibitory impulses (i-events) in a Poisson stream with rate μ.

c) An e-event results in an EPSP of size α and an i-event in an IPSP of size β.

d) When $X(t)$ crosses the threshold level K, the neuron fires and $X(t)$ returns to x_0 (as stated in a)) and the firing process repeats itself.

e) When there is no input the membrane potential decays to the rest level x_r exponentially with time constant τ. (In the equations to follow, x_r is set equal to zero without loss of generality).

f) The changes α and β are small compared to $K-x_0$ and $K-x_r$.

With the above assumptions it can be shown that (6.1.29) and (6.1.30) become

$$a(x,t) = c - \frac{x}{\tau} \qquad (6.2.1)$$

and

$$b(x,t) = \sigma^2 \qquad (6.2.2)$$

where

$$c = \lambda\alpha - \mu\beta \qquad (6.2.3)$$

and

$$\sigma^2 = \lambda\alpha^2 + \mu\beta^2. \qquad (6.2.4)$$

The diffusion equation (6.1.33) thus takes the form

$$f_t = - \frac{\partial}{\partial x} [c - \frac{x}{\tau}) f] + \frac{1}{2} \sigma^2 f_{xx} \qquad (6.2.5)$$

where $f(x,t \mid x_0, t_0)$ is written f for convenience.

Since the e-and i-sequences are both Poisson, the sequence of firings is a renewal process. Therefore it is completely described by the first passage time pdf.

6.2.1 Model 6.1 (Roy and Smith, 1969; Griffith, 1971; Capocelli and Ricciardi, 1973)

This model is the basic diffusion model described in the last paragraph. Roy and Smith (1969) started with the one-dimensional equation with drift (6.1.22) and modified it by writing $c - \frac{x}{\tau}$ in place of c. This was heuristically justified on the grounds that c is the drift of the membrane potential in the absence of decay while the rate of change of the potential due to exponential decay alone is

$$\frac{\partial x}{\partial t} = - \frac{x}{\tau}, \qquad (6.2.6)$$

so that the rate of change due to both effects is $c - \frac{x}{\tau}$. Equation (6.2.5) was also derived by Griffith (1971) directly using Taylor series expansions with appropriate approximations. Griffith attempted to remove the approximation of small changes in the membrane potential (see assumption (f)) and considered the EPSPs and IPSPs to be random variables (see p.11) with identical distribution $A(\alpha)$ and $B(\beta)$ respectively. The resulting equation is

$$f_t = \frac{x}{\tau} f_x - (\lambda+\mu) f + \frac{1}{\tau} f$$

$$+ \lambda \int_0^\infty f(x-\alpha, t \mid x_0, t_0)\, A(\alpha)\, d\alpha$$

$$+ \mu \int_0^\infty f(x+\beta, t \mid x_0, t_0)\, B(\beta)\, d\beta. \tag{6.2.7}$$

This, however, is an intractable equation. Therefore this line of thought will not be pursued further. Solutions of (6.2.5) only will be considered here.

Using a series of transformations the initial value problem corresponding to (6.2.5) can be solved the same way as (6.1.22). This solution in turn can be used to arrive at an expression for the Laplace transform of the first passage time distribution. This, however, will not be done here because the result is not very useful. (See Griffith (1971, p. 66) for the details). A more fruitful way of arriving at the transform of the first passage time distribution will be considered here. This result too is not in closed form but leads to more tangible results.

Let T be the r.v. of the first passage time of the membrane potential $X(t)$ to the threshold K and $g(t \mid x_0, 0)$ its pdf, where the diffusion process is assumed to start at $x_0 (< K)$ at $t_0=0$.

<u>Theorem 6.2.1</u> The Laplace transform $g^*(s \mid x_0, 0)$ of the first passage time pdf $g(t \mid x_0, 0)$ in the diffusion model with exponential decay in the absence of input impulses satisfies the ordinary differential equation

$$\frac{\sigma^2}{2} \frac{d^2 g^*}{dx_0^2} + \left(c - \frac{x_0}{\tau}\right) \frac{dg^*}{dx_0} - s g^* = 0. \tag{6.2.8}$$

Proof: Consider the diffusion process $X(t)$ with the threshold K as a barrier. Define a process $Y(t)$ such that

$$Y(t) = X(t) < K \text{ in } (0,t).$$

Let

$$P(x,t \mid x_0,0) = \text{Prob}\,[Y(t) < x \mid Y(0) = x_0]. \qquad (6.2.9)$$

Thus

$$P(K,t \mid x_0,0) = \text{prob}\left\{ \text{the membrane potential has not crossed the threshold } K \text{ in } (0,t)\right\}. \qquad (6.2.10)$$

It then follows that

$$g(t \mid x_0,0) = -\,P_t(K,t \mid x_0,0). \qquad (6.2.11)$$

Since $P(K,t \mid x_0,0)$ satisfies (6.1.32) it can be easily seen that

$$g_t(t \mid x_0,0) = (c - \frac{x_0}{\tau})\, g_{x_0}(t \mid x_0,0) + \frac{1}{2}\sigma^2\, g_{x_0x_0}(t \mid x_0,0). \qquad (6.2.12)$$

Taking Laplace transforms of both sides and remembering that $g(0 \mid x_0,0) = 0$, (6.2.8) is obtained. This proves Theorem 6.2.1.

The ordinary differential equation (6.2 . 8) is now to be solved with the boundary conditions

$$g^*(s \mid K,0) = 1 \qquad (6.2.13)$$

and

$$g^*(s \mid -\infty,0) = 1 \qquad (6.2.14)$$

or, more conveniently,

$$\lim_{K' \to \infty} g^*(s \mid -K',0) = 1. \qquad (6.2.15)$$

These two conditions imply that if the initial state is either of the two barriers K and -K', the absorption probability is 1. Equation (6.2.8) is solved by the usual methods. See, for example, Langer (1954). Since the methods are standard only the results will be stated below. The transform of the first passage time distribution, the mean firing interval $\bar{T}$ and some special cases are all expressed in terms of the confluent hypergeometric function and its asymptotic expansion (see Abramowitz and Stegun, 1965, p. 503).

<u>Theorem 6.2.2</u> The **Laplace** transform of the first passage time distribution in the basic diffusion model with exponential decay is given by

$$g^*(s \mid x_0, 0) = \frac{\Psi\left(\frac{s\tau}{2}, \frac{1}{2}, \frac{(c\tau - x_0)^2}{\sigma^2\tau}\right)}{\Psi\left(\frac{s\tau}{2}, \frac{1}{2}, \frac{(c\tau - K)^2}{\sigma^2\tau}\right)}, \tag{6.2.16}$$

where Ψ is the confluent hypergeometric function of the second kind.

The usefulness of the transform lies in the ease with which the moments can be found out by differentiation, although the transform itself is of little help because of the complexity of inversion.

<u>Theorem 6.2.3</u> The mean firing interval in the basic diffusion model with exponential decay is given by

$$\bar{T} = \frac{\tau}{2}\left[\sum_{k=0}^{\infty} \frac{2^{k+1}\left(\frac{c-K/\tau}{\sigma/\sqrt{\tau}}\right)^{2k+2}}{(2k+1)!!\,(k+1)}\right.$$

$$- \sum_{k=0}^{\infty} \frac{\left(\frac{c - x_0/\tau}{\sigma/\sqrt{\tau}}\right)^{2k+2} 2^{k+1}}{(2k+1)!!\,(k+1)}$$

$$+ 2\sqrt{\pi}\left\{\frac{c-x_0/\tau}{\sigma/\sqrt{\tau}}\,\Phi\left(1/2, 3/2, \left(\frac{c-x_0/\tau}{\sigma/\sqrt{\tau}}\right)^2\right)\right.$$

$$\left.\left. - \frac{c-K/\tau}{\sigma/\sqrt{\tau}}\,\Phi\left(1/2, 3/2, \left(\frac{c-x_0/\tau}{\sigma/\sqrt{\tau}}\right)^2\right)\right\}\right] \tag{6.2.17}$$

where Φ is the confluent hypergeometric function.

In order to relate the output firing sequence to the input to the neuron the following quantities are introduced. Let $\alpha=\beta=1$. Thus the EPSP and IPSP are both set equal to unity. The mean firing rate

$$\lambda_0 = 1/\bar{T}. \tag{6.2.18}$$

The drift ratio

$$d = (\lambda - \mu)/(\lambda + \mu). \tag{6.2.19}$$

From these, the drift parameter

$$c = d(\lambda + \mu) \tag{6.2.20}$$

and the diffusion coefficient

$$\sigma^2 = (\lambda + \mu). \tag{6.2.21}$$

Before presenting the input-output graphs the asymptotic behaviour of $\bar{T}$ for large $(\lambda + \mu)$ is given by two corollaries.

<u>Corollary 1</u> As $(\lambda + \mu) \to \infty$,

$$\bar{T} \to \tau \ln\left(\frac{c\tau - x_0}{c\tau - K}\right). \tag{6.2.22}$$

<u>Corollary 2</u> Further if $c \gg K/\tau$, i.e., the decay becomes ineffective,

$$\lambda_0 \to (\lambda - \mu)/K. \tag{6.2.23}$$

The same results are obtained when $\sigma^2 = 0$, i.e., when the input sequences are deterministic.

Graphs relating λ_0 and $(\lambda + \mu)\tau/K$ are presented in Figure 6.2.1 for different values of K, τ and d. The right extreme curve is for the asymptotic result given by Corollary 1. The dashed line corresponds to the case given in Corollary 2. For the details, see Roy and Smith (1969).

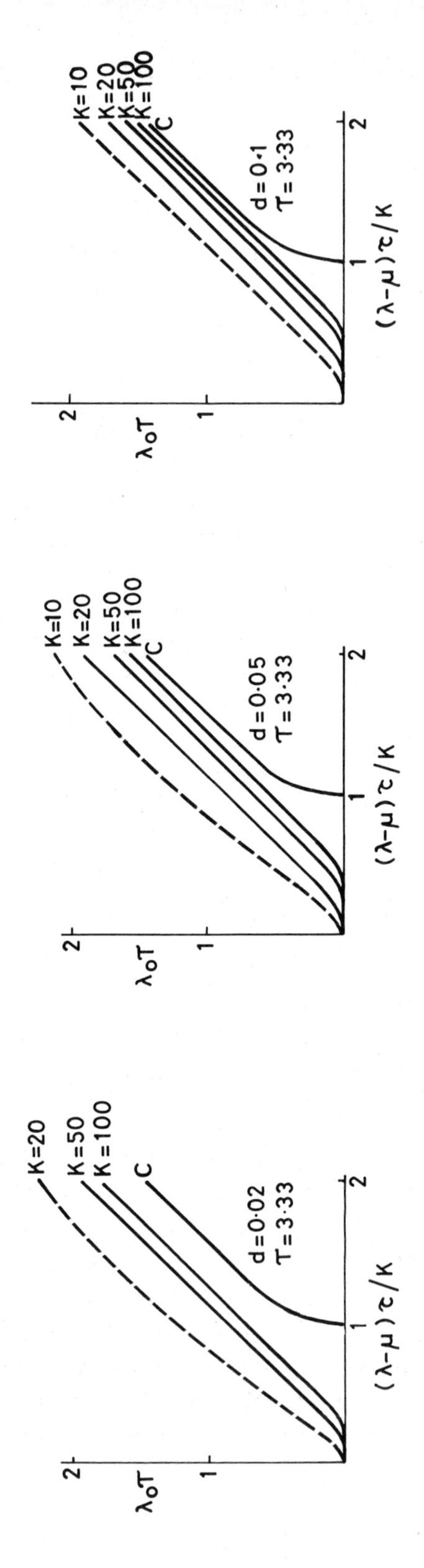

FIG.6.2.1. INPUT-OUTPUT CURVES-MODEL 6·1 (Roy and Smith,1969)

6.2.2 Model 6.2 - Refractoriness in a diffusion model (Clay and Goel, 1973)

In Section 1.1.1, the refractory property of a neuron after a firing was discussed. There are two periods of inactivity following spike generation. The first is the absolute refractory period (1 to 2 msec) and the second is the relative refractory period (~ 5 m sec). In the former, the threshold is infinite and in the latter the threshold returns from infinity to the normal level (i.e., the level when the neuron is not active). Since the absolute refractory period is more or less constant this is easily taken care of in the diffusion model because the input sequences are Poisson streams. The relative refractory period is more difficult to model. The return of the threshold to the normal level has been considered in some models by assuming appropriate functions of time for the threshold. Two of them are the exponential threshold and the Hagiwara threshold. These are respectively given by

$$K(t) = K + K_1 e^{-t/\tau_1}, \quad K_1 \gg K \tag{6.2.24}$$

and

$$K(t) = K e^{\tau_2/t}, \quad \tau_2 > 0 ; \tag{6.2.25}$$

in both cases

$$K(t) \rightarrow K \quad \text{as} \quad t \rightarrow \infty . \tag{6.2.26}$$

Clay and Goel (1973) have obtained solutions of the diffusion equation with these threshold functions for special cases. Their results are discussed below.

Consider a diffusion model with the basic features given in Section 6.2. In addition to these, let the neuron membrane potential after a firing stay at x_0 till time $t_0 + \tau_A$, where τ_A is the absolute refractory period. After $t_0 + \tau_A = t_1$ let the threshold $K(t)$ decay according to the law $K(t-t_1)$ where $K(\cdot)$ is given by one of the equations (6.2.24) and (6.2.25). Since the inputs are Poisson sequences the absolute refractory period τ_A can easily be included in the final results. Hence t_1 will be set equal to zero for convenience. Model 6.2 therefore reduces to the solution of the boundary value problem

$$f_t = -\frac{\partial}{\partial x}[c - \frac{x}{\tau}) f] + \frac{1}{2}\sigma^2 f_{xx}$$

with

$$f(x = K(t), t \mid x_0, 0) = 0. \tag{6.2.27}$$

The following theorem for the special case $\tau_1 = \tau$ **is stated** without proof.

<u>Theorem 6.2.4</u> The first passage time distribution $g(K(t),t \mid x_0,0)$ in the diffusion model with the threshold varying as

$$K(t) = K + K_1 e^{-t/\tau}$$

is given by the Laplace transform

$$g^*(s \mid x_0,0) = \int_0^\infty g(K(t),t \mid x_0,0)\, e^{-st}\, dt$$

$$= \frac{D_{-s}(\eta)\, e^{(\eta^2-\eta_0^2)/4}}{D_{-s}(\eta_0)} \tag{6.2.28}$$

where $D_{-s}(\cdot)$ is the parabolic cylinder function (Abramowitz and Stegun (196 , p.686))

$$\eta = \frac{\sqrt{2}(x_0 - c\tau - K_1)}{\sigma\sqrt{\tau}}, \tag{6.2.29}$$

and

$$\eta_0 = \frac{\sqrt{2}(c\tau - K)}{\sigma\sqrt{\tau}}. \tag{6.2.30}$$

<u>Corollary 1</u> When $c = K/\tau$, the first passage time is given by

$$g(K + K_1 e^{-t/\tau}, t \mid x_0, 0)$$

$$= \frac{2(K_1-x_0+c\tau)^2}{\sigma\tau\sqrt{\pi\tau}} \frac{e^{2t/\tau}}{(e^{2t/\tau}-1)^{3/2}}$$

$$\exp\left[- \frac{K_1-x_0+c\tau}{\sigma^2\tau\,(e^{2t/\tau}-1)}\right]. \tag{6.2.31}$$

From (6.2.28) the mean firing rate can be obtained as

$$\lambda_o = 1/(\bar{T} + \tau_A), \tag{6.2.32}$$

where the absolute refractory period τ_A has been included and $\bar{T}$ is

$$\bar{T} = \frac{\tau}{2}\left[\sum_{k=0}^{\infty} \frac{2^{k+1}\,(\eta_0^{2k+2} - \eta^{2k+2})}{(2k+1)!!\,(k+1)}\right.$$

$$+ 2\sqrt{\pi}\left\{\eta_0\,\Phi(1/2,3/2,\eta_0^2) - \right.$$

$$\left.\left.\eta\,\Phi(1/2,3/2,\eta_0^2)\right\}\right] \tag{6.2.33}$$

where Φ is the confluent hypergeometric function.

For threshold variations other than the one considered above, i.e., $K(t) = K + K_1 e^{-t/\tau}$, the first passage time problem can be solved if more approximations are made. A method of solution is considered here. Assume $K(t)$ to be constant (=K, say), but all other processes are slowed down by the function

$$i(t) = \frac{K}{K(t)}. \tag{6.2.34}$$

Thus

$$\lambda \longrightarrow \lambda\, i(t) \tag{6.2.35}$$

$$\mu \longrightarrow \mu\, i(t) \tag{6.2.36}$$

$$\tau \longrightarrow \tau/i(t), \tag{6.2.37}$$

Then (6.2.27) becomes

$$\frac{\partial f}{\partial I} = - \frac{\partial}{\partial x} [(c - \frac{x}{\tau}) f] + \frac{1}{2} \sigma^2 \frac{\partial^2 f}{\partial x^2} , \qquad (6.2.38)$$

where

$$I(t) = \int_0^t i(u)\, du. \qquad (6.2.39)$$

with the boundary condition

$$f[K, I(t) \mid x_0, I(t_0)] = 0. \qquad (6.2.40)$$

The problem has now reduced to a first passage time problem with constant threshold and the solution of this is given in Section 6.2.1. Make the transformation

$$I(t) \longrightarrow t , \qquad (6.2.41)$$

$$g(K, I(t) \mid x_0, I(t_0)) \longrightarrow$$

$$\frac{1}{i(t)} g\,(K(t), t \mid x_0, t_0). \qquad (6.2.42)$$

This gives the solution of the varying threshold neuron model.

For the cat spinal motoneuron , $\tau = 5$ msec, the PSP $\alpha = \beta \sim 0.25$ mV, the threshold level $K \sim 10$ mV, the absolute refractory period $\tau_A \sim 1$ to 2 msec, and the relative refractory period $\tau_R \sim 4$ msec (corresponding to the time the neuron returns to the level at which the effectiveness for generating a spike is 90%). Thus in (6.2.34)

$$i(t) = 0.9.$$

For the Hagiwara threshold (6.2.25) putting

$$i(\tau_R) = 0.9 = e^{-K_2/\tau_R}$$

one gets

$$K_2 \approx 0.1\ \tau_R.$$

For the exponential threshold (6.2.24), set

$$i(\tau_R) = 0.9 = \left(1 + \frac{K_1}{K} e^{-\tau_R/\gamma}\right)^{-1}$$

in which one of K_1 and γ has to be assumed. Since $K_1 \gg K$, it is reasonable to assume

$$K_1 = 100\ K.$$

(Because the input-output relation is insensitive to the magnitude of K_1 for large K_1/K, it does not matter what K_1/K is provided it is large enough). Using these numerical values and the solutions of the first passage time problem given earlier in this section, Clay and Goel (1973) presented input-output graphs **relating** $\lambda_0\tau$ and $(\lambda - \mu)\tau/K$ for the three cases (i) constant threshold (ii) Hagiwara threshold (iii) exponential threshold. These are given in Figure 6.2.2, which shows that the model is not very sensitive

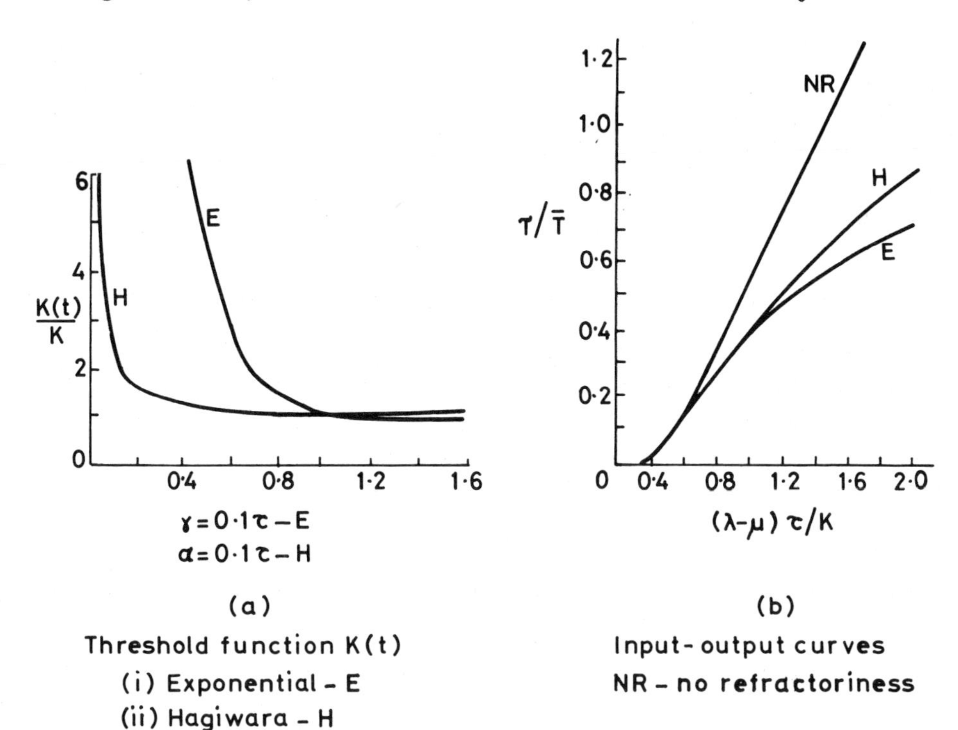

(a)
Threshold function K(t)
(i) Exponential - E
(ii) Hagiwara - H

(b)
Input-output curves
NR - no refractoriness

FIG. 6.2.2. INPUT-OUTPUT CURVES-MODEL 6·2
(Clay and Goel, 1973)

to the form of variation of the threshold.

For a complete analysis of the model presented in this section see Clay and Goel (1973).

6.3 Discussion

The diffusion model is a useful description of the firing process when the input sequences (excitatory and inhibitory) are both Poisson. It is applicable to certain neurons in which the threshold is around 40 to 50 mV and the PSPs are approximately 3 or 4 mV (Eccles, 1964). The Poisson nature of the input sequences makes it easy to incorporate refractoriness in the model because the origin is arbitrary in a Poisson process. Also thresholds that vary with time can be considered in the model. A striking feature of the diffusion model is that the threshold as a function of time does not affect the differential equation but only occurs as a time-dependent boundary condition. Though the analytic solution of such a boundary value problem may be difficult, since the basic equation is the same the complexity of the numerical solution is not increased greatly. The exponential decay of the membrane potential is brought into the model in a natural way and this gives the model a good amount of realism. It is interesting to note here that if decay is not included in the diffusion model, the total probability of the membrane potential reaching the threshold is less than 1 (see Griffith, 1971, p.61; Cox and Miller, 1965, p.211) when $c < 0$. Hence in a sense the decay of the membrane potential becomes a mathematical necessity.

The diffusion model has severe limitations. If the input sequences are not Poisson, the model is inapplicable. Moreover, the Poisson assumption makes modelling of dependent interaction of sequences as occurs in feedback impossible. Also the size of the PSPs is not small in many kinds of neurons, and in addition may be a random variable (see p.11).

CHAPTER 7

COUNTER MODELS

Consider particles arriving at random intervals of time at a counter. The counter registers or clicks on the arrival of a particle. A practical counter is for sometime thereafter unable to register an arrival. This period of inactivity is called <u>dead time</u>: the counter goes 'dead'. Immediately after this dead time, the counter is able to register arrivals. The similarity of the dead time effect with refractoriness in a neuron is at once obvious. This has motivated the use of counter theory in modelling neuron spike trains.

Counters that are commonly used in particle detectors form an interesting class of renewal processes. In this chapter, counter theory is applied to the modelling of spontaneous activity in single neurons. First, the theory of counters is discussed briefly in Section 7.1, following Cox (1962) and Srinivasan and Rajamannar (1970c). In Sections 7.2 and 7.3 counter models based on the basic deletion model (Model 5.1) are presented. In Section 7.3 a non-Markovian model of a neuron with threshold behaviour is studied.

7.1 Theory of counters

Let particles arrive at a counter in a stationary sequence that forms a renewal process with interval pdf $\varphi(\cdot)$. For convenience call the event of the arrival of a particle a primary event. If a primary event is registered, it is followed by a dead time T during which no event is registered. There are two types of counters depending on the behaviour of the counter during the dead time —

<u>Type I</u>: An unregistered event does not lock the counter.

<u>Type II</u>: Each event registered or unregistered gives rise to a dead time.

The primary aim in counter studies is to obtain the characteristics of the sequence of registered events. Since the renewal nature of the primary events implies that the registered events form a renewal process, the process is completely characterised by the pdf $f(\cdot)$ of the interval between two successive registered events. However it may not always be easy to derive an expression for $f(\cdot)$; at least in some cases it may be easier to obtain the renewal density of the registered events.

Theorem 7.1.1 In a Type I counter if the primary events form a renewal process with pdf $\varphi(\cdot)$ and the dead times are all iid r.v.s with pdf $\eta(\cdot)$, then the pdf $f(\cdot)$ of the registered events is given by

$$f(t) = \varphi(t)\, R(t) + \int_0^t h_e(u)\, \varphi(t-u)\, du \int_u^t \eta(v)\, dv \tag{7.1.1}$$

where

$$R(t) = \int_0^t \eta(v)\, dv \tag{7.1.2}$$

and $h_e(\cdot)$ is the renewal density of the primary events and is related to $\varphi(\cdot)$ by the renewal equation

$$h_e(t) = \varphi(t) + \int_0^t h_e(u)\, \varphi(t-u)\, du. \tag{7.1.3}$$

The theorem is proved using the renewal nature of the primary events.

Corollary 1. If the primary events form a Poisson process with parameter λ,

$$f^*(s) = \frac{\lambda\, \eta^*(s)}{\lambda + s}. \tag{7.1.4}$$

If, in addition, the dead time is a constant a,

$$f^*(s) = \frac{\lambda e^{-as}}{\lambda + s}. \tag{7.1.5}$$

From (7.1.5) follows

<u>Corollary 2</u>. The renewal density $h(\cdot)$ of the registered events has the Laplace transform

$$h^*(s) = \frac{\lambda}{(\lambda+s)e^{as} - \lambda}, \tag{7.1.6}$$

with the asymptotic result (obtained by using the Tauberian theorem)

$$h(t) \longrightarrow \lambda/(a\lambda+1), \quad \text{as } t \longrightarrow \infty. \tag{7.1.7}$$

<u>Theorem 7.1.2</u>. In a Type II counter if the primary events form a renewal process with interval pdf $\varphi(\cdot)$ and the dead times (due to every event, registered or unregistered) are iid r.v.s with pdf $\eta(\cdot)$, the renewal density $h(\cdot)$ of the registered events is given by

$$h(t) = \varphi(t)\,R(t) + \sum_{n=1}^{\infty} \int_0^t dt_1 \int_{t_1}^t dt_2 \cdots \int_{t_{n-1}}^t \varphi(t_1)\varphi(t_2 - t_1)$$

$$\cdots \varphi(t - t_n)\, R(t)R(t - t_1)\cdots R(t - t_n)\, dt_n\,. \tag{7.1.8}$$

This theorem is proved by observing that for an event to occur in $(t, t+dt)$ it must be beyond the dead times of all the earlier events.

<u>Theorem 7.1.3</u>. In a Type II counter if the primary events are Poisson with parameter λ, a registered event is followed by a constant dead time of length $\underline{a}$ and an unregistered event by a constant dead time of length $\underline{b}$, $\underline{a} > \underline{b}$, then the renewal density of the registered events has the Laplace transform

$$h^*(s) = \frac{\lambda}{se^{as+b} + \lambda e^{(a-b)s} - \lambda}. \tag{7.1.9}$$

For the proof, see Ramakrishnan and Mathews (1953).

<u>Corollary</u>. The asymptotic value of $h(t)$ is given by

$$h(t) \longrightarrow \frac{\lambda}{\lambda(a-b) + e^{b\lambda}}, \tag{7.1.10}$$

which reduces to (7.1.7) when $b=0$.

In the above discussion it has been assumed that during the dead time any number of primary events remain unregistered. It may be more appropriate to consider a modified dead time which lasts only upto its lifetime (determined by its probability law) or upto the first primary event following the registered event, whichever is earlier. With this modification, the two types of counters are called 'modified Type I' and 'modified Type II' counters respectively.

Theorem 7.1.4. In a modified Type I counter, the interval pdf of the registered events is given by

$$f(t) = \varphi(t)\,R(t) + \int_0^t \varphi(u)\,\varphi(t-u)\,[1-R(u)]\,du. \tag{7.1.11}$$

Theorem 7.1.5. In a modified Type II counter, the renewal density of the registered events $h(\cdot)$ is given by

$$h(t) = \varphi(t)R(t) + \int_0^t h_e(u)R(t-u)\varphi(t-u)du, \tag{7.1.12}$$

where $h_e(\cdot)$ is the renewal density of the primary events.

The proofs of Theorems 7.1.4 and 7.1.5 are fairly simple and are omitted.

Since the primary events arriving during a dead time themselves result in dead times, the effective dead time increases. The following theorem relating to this effective dead time will be needed later in one of the models.

Theorem 7.1.6 . The pdf $\theta(\cdot)$ of the effective dead time in a modified Type II counter when the primary events are Poisson with parameter λ has the Laplace transform

$$\theta^*(s) = \frac{(s+\lambda)\eta^*(s+\lambda)}{s+\lambda\eta^*(s+\lambda)} . \tag{7.1.13}$$

Proof. It is easy to see that the interval pdf $f(\cdot)$ of the registered events in a modified Type II counter is related to the dead time distribution $\theta(\cdot)$ through

$$f(t) = \int_0^t \theta(u)\,\lambda\,e^{-\lambda(t-u)}\,du, \tag{7.1.14}$$

from which

$$\theta^*(s) = f^*(s)(s+\lambda)/\lambda . \tag{7.1.15}$$

Now, $f^*(s)$ is obtained from the Laplace transform of (7.1.12) in the usual way. Using this in (7.1.15) and simplifying, equation (7.1.13) results.

Compare (7.1.13) with (8.3.41).

In view of the importance of the Poisson process in deriving simplified models, it is worthwhile finding the conditions under which the registered events form a Poisson process with parameter, say, ρ . These are obtained by writing

$$f(t) = \rho e^{-\rho t} \tag{7.1.16}$$

in (7.1.11), and

$$h(t) = \rho \tag{7.1.17}$$

in (7.1.12).

In the modified Type I counter the resulting integral equation is nonlinear. If however the dead time distribution is exponential, i.e.,

$$\eta(v) = ke^{-kv}, \tag{7.1.18}$$

then it is easy to arrive at

<u>Theorem 7.1.7</u> . In the modified Type I counter if the dead time is exponentially distributed with parameter k and the registered events form a Poisson process with parameter ρ , then the Laplace transform of the pdf of the interval between two primary events satisfies

$$\varphi^*(s) + \varphi^*(s)\,\varphi^*(k+s) - \varphi^*(k+s) = \rho/(\rho+s). \tag{7.1.19}$$

In the case of the modified Type II counter, the distribution $\varphi(t)$ can be obtained explicitly.

Theorem 7.1.8 . In the modified Type II counter, the registered events form a Poisson process with parameter ρ if the primary events form a renewal process with interval pdf $\varphi(\cdot)$ satisfying

$$\int_0^t \varphi(t')\, dt' = \int_0^t \exp\left[-\int_{t'}^t \frac{\rho\, du}{1 - e^{-ku}}\right] \frac{\rho\, dt'}{1 - e^{-kt'}} \qquad (7.1.20)$$

and the dead times are iid with pdf ke^{-kt}.

The above discussion illustrates the methods of counter theory. These can be used fruitfully in neuron modelling particularly when considering refractoriness. Another use of counter studies is in bringing into the neuron model the effectiveness of inhibition. Thus the deletion models of Chapter 5 are extended in many ways: the pre-synaptic inhibition effect of an i-event is assumed to last for a period of time T after the i-event and this is similar to the dead time property. In Section 7.2, the deletion model with independent interaction of e- and i-sequences of Section 5.1 is extended by including dead time effects due to i-events. In Section 7.3 the same is done for the dependent interaction models of Section 5.2. In Section 7.4 the cumulative effect of registered events is considered in a non-Markovian model with threshold.

7.2 Counter model extensions of deletion models with independent interaction of e- and i-events (cf. Section 5.1)

Pre-synaptic inhibition has been modelled in Chapter 5. In the basic deletion model it is assumed that the i-event is effective for an infinite length of time. This is not very realistic because the inhibitory effect is due to chemical reactions that are reversible. Hence it is more meaningful to consider the i-event 'mortal'. Based on this concept several extensions have been proposed. Thus Coleman and Gastwirth (1969) proposed three models which are extensions of Model 5.1. In the first, an i-event is followed by a constant dead time T during which no e-event registers. In the second, the dead time T is assumed to be a r.v. that is exponentially distributed. In the third, all e-events (rather than just the first) arriving within the dead time T are deleted. Srinivasan and Rajamannar (1970c) extended some of these results. Similar extensions have been considered by Pooi Ah Hin (1974). In all these, the i-events are assumed to form a Poisson process, for otherwise the resulting process of undeleted e-events is non-renewal and is considerably difficult to study. Some of these models are described below.

7.2.1 Model 7.1 (Coleman and Gastwirth, 1969) - Extension of Model 5.1

This model has the following features:

1) The e-events form a renewal process with interval pdf $\varphi(\cdot)$ and the i-events a Poisson process with parameter μ.
2) An i-event deletes the first e-event only if the e-event arrives within a time T after the i-event.
3) Undeleted e-events (or r-events) form the output in the model.

Theorem 7.2.1 . In Model 7.1, the Laplace transform of the interval between two successive r-events is given by

$$f^*(s) = \frac{\varphi_T^*(s)}{1 - \varphi^*(s) + \varphi_T^*(s)}, \tag{7.2.1}$$

where

$$\varphi_T^*(s) = \int_0^T e^{-(s+\mu)t}\,\varphi(t)\,dt + e^{-\mu T}\int_T^\infty e^{-st}\,\varphi(t)\,dt. \tag{7.2.2}$$

The proof is simple and is omitted.

7.2.2 Model 7.2 (Srinivasan and Rajamannar, 1970c ; Coleman and Gastwirth, 1969) - Extension of Model 5.1

This model has the same properties as Model 7.1 with the only difference being

4) T is a r.v. with pdf $\eta(\cdot)$.

To obtain the pdf of the interval between two registered events a number of subsidiary functions are introduced. First it is recognised that the i-events behave like the arrivals in a Type II counter wherein the primary events that are not registered themselves lead to a dead time. Thus if an i-event arrives at time $t = 0$, it creates a dead time T during which other i-events may arrive and thus effectively increase the dead time to a larger value, say, D. Any e-event arriving in (0,D) is deleted. The system may thus be presumed to be 'dead' in (0,D). The dead time ends with a deletion or because of the probability law; the system then becomes 'free' again.

Define

$\pi_0(t)$ = Prob [system is 'free' at time t | it is 'free' at $t = 0$], (7.2.3)

$\pi_1(t)$ = Prob [system is 'free' at time t | it goes 'dead' at $t = 0$], (7.2.4)

$\omega_0(t)$ = Prob [system is 'dead' at time t | it becomes 'free' at $t = 0$], (7.2.5)

$\omega_1(t)$ = Prob [system is 'dead' at time t | it goes 'dead' at $t = 0$]. (7.2.6)

Then the following results are easily obtained:

$$\pi_0(t) = e^{-\mu t} + \int_0^t e^{-\mu u} \mu \, du \, \pi_1(t-u), \tag{7.2.7}$$

$$\pi_1(t) = \int_0^t \Theta(u) \, \pi_0(t-u) \, du, \tag{7.2.8}$$

$$\omega_0(t) = 1 - \pi_0(t), \tag{7.2.9}$$

$$\omega_1(t) = 1 - \pi_1(t), \tag{7.2.10}$$

where $\Theta(\cdot)$ is the pdf of the total dead time D in a Type II counter.

Lemma 7.2.1 . If

$$q(t) = \lim_{\Delta, \Delta' \to 0} \text{Prob [an unregistered e-event occurs in } (t, t+\Delta) \text{ and no r-event in } (0,t) \mid \text{an r-event in } (-\Delta', 0)]/ \tag{7.2.11}$$

then $q(\cdot)$ satisfies

$$q(t) = \varphi(t)\, \omega_0(t) + \int_0^t q(u)\, \omega_0(t-u)\, \varphi(t-u)\, du. \qquad (7.2.12)$$

Using the above lemma and the subsidiary functions given by (7.2.7) to (7.2.10), one can arrive at

Theorem 7.2.2 . The pdf $f(\cdot)$ of the interval between two successive r-events in Model 7.2 is given by

$$f(t) = \varphi(t)\, \pi_0(t) + \int_0^t q(u)\, \pi_0(t-u)\, \varphi(t-u)\, du. \qquad (7.2.13)$$

The condition for the r-events to form a Poisson process with parameter ρ is given by the following theorem which is stated without proof.

Theorem 7.2.3 . The r-events in Model 7.2 form a Poisson process with parameter ρ if the e-events form a renewal process with pdf $\varphi(\cdot)$ satisfying

$$\varphi(t) = \frac{\rho}{\pi_0(t)} \exp\left[1 - \int_0^t \frac{\rho\, du}{\pi_0(u)} \right] \qquad (7.2.14)$$

and the i-events form a Poisson stream with parameter ρ and have a dead time distribution $\eta(\cdot)$.

7.2.3 Model 7.3 (Coleman and Gastwirth, 1969)

This is the same as Model 7.2 with the following modifications:

2) One or more i-events delete all e-events arriving within the dead time T.
4) The e-events are Poisson with parameter λ .

Theorem 7.2.4 . The Laplace transform of the pdf of the interval between two successive r-events in Model 7.3 is given by

$$f^*(s) = \left\{ 1 + [\lambda\, Z^*(s)]^{-1} \right\}^{-1}, \qquad (7.2.15)$$

where

$$Z(t) = \exp [- \mu \int_0^t dx \int_x^\infty \eta(u) \, du]. \qquad (7.2.16)$$

Coleman and Gastwirth have further extended the model by using the property of the modified Type II counter (Section 7.1). For the details, see Coleman and Gastwirth (1969).

7.3 Counter model extensions of deletion models with dependent interaction of e- and i-events (cf. Section 5.2)

In Section 5.2 a number of deletion models were discussed in which one type of event gives rise to a sequence of the other type. These models are modified here (as was done with the independent interaction models in Section 7.2) by introducing a dead time effect due to the i-events. Srinivasan and Rajamannar (1970c) have assumed that the i-events are effective only when they arrive within time T of the first induced i-event. Osaki (1971) has extended some of these models. Since the approach is very similar to that in the last section, the discussion **here** is brief.

7.3.1 Model 7.4 (Srinivasan and Rajamannar, 1970c) - Extension of Model 5.3

The model has the following features:

1) The e-events form a renewal process with pdf $\varphi(\cdot)$.
2) Each e-event gives rise to a sequence of i-events (the time to the first i-event after the e-event having pdf $\psi(\cdot)$ and the sequence of i-events forming a renewal process with pdf $\psi(\cdot)$) and stops the sequence of i-events generated by the previous e-event.
3) An i-event deletes the next e-event only if the i-event arrives within a time T after the first i-event in the sequence of i-events, T being a r.v. with pdf $\eta(\cdot)$.
4) Undeleted e-events (r-events) form the output in the model.

Theorem 7.3.1. The renewal density $h(\cdot)$ of the r-events in Model 7.4 is given by

$$h(t) = \varphi(t) [\int_t^\infty \psi(u) \, du + \int_0^t du \int_0^u \psi(v) \, \eta(u-v) dv]$$

$$+ \int_0^t h_e(u)\, du\, \varphi(t-u) \left[\int_{t-u}^{\infty} \psi(v)\, dv + \int_{t-u}^{\infty} dv \int_0^v \psi(w)\, \eta(v-w) dw \right]. \tag{7.3.1}$$

Theorem 7.3.2 . The r-events in Model 7.4 constitute a Poisson process with parameter ρ if the e-events form a renewal process with pdf $\varphi(\cdot)$ given by

$$\varphi(t) = \exp \left[- \int_0^t \frac{\rho\, du}{L(u)} \right] \frac{\rho}{L(t)} , \tag{7.3.2}$$

where

$$L(t) = \int_t^{\infty} \psi(u)\, du + \int_0^t du \int_0^u \psi(v)\, \eta(u-v)\, dv \tag{7.3.3}$$

and $\eta(\cdot)$ is the pdf of the dead time following the first i-event generated by an e-event.

7.3.2 Model 7.5 (Srinivasan and Rajamannar, 1970c) - Extension of Model 5.4

This model has the same features as Model 7.4 with the only modification in

4) Not every e-event gives rise to a sequence of i-events. Only an r-event has this effect.

Theorem 7.3.3 . The pdf $f(\cdot)$ of the interval between two successive r-events in Model 7.5 is given by

$$f(t) = \varphi(t) \left[\int_t^{\infty} \psi(u)\, du + \int_0^t du \int_0^u \psi(v)\, \eta(u-v) dv \right]$$

$$+ \int_0^t \varphi(u)\, \varphi(t-u)\, du \int_0^u \psi(v)\, dv \int_{u-v}^{\infty} \eta(w)\, dw \quad . \tag{7.3.4}$$

An attempt to find out the condition for the r-events to form a Poisson process results in a nonlinear equation. Hence this is not pursued further.

7.4 Counter models with threshold behaviour

In the previous sections, the counter models discussed have a very serious defect, viz., that they do not consider the integrating effect that the membrane has over successive r-events. An obvious way to overcome this is to use the counting function $N(t)$ - see p. 27 - and build a first passage time model. Thus if the r-events form a renewal process, the first passage time to a threshold level N is the time to the N-th r-event after the origin, and the pdf of this completely describes the process. A more general model in which the cumulative response of the registered events in a counter with the response due to a single registered event being given by a response function $g(\cdot)$ crosses a threshold level K has been studied by Srinivasan and Vasudevan (1969). This is discussed briefly below.

7.4.1 Model 7.6 (Srinivasan and Vasudevan, 1969)

The model has the following features:

1) The registered events in a counter form a renewal process with pdf $f(\cdot)$.
2) The post-synaptic potential due to an r-event on the membrane is given by $g(t-t_i)$ where t_i is the time of occurrence of the i-th r-event. The membrane potential $X(t)$ is the sum of all these potentials.
3) When the total response due to the r-events crosses the threshold level K, the neuron fires and is reset to the rest level (zero).

Notice that $g(\cdot)$ does not reflect the decay of the total membrane potential $X(t)$ but the effect of an induced EPSP at any time after it is generated by an r-event. In a sense this is a kind of memory effect. Hence the model may be considered a <u>non-Markovian</u> model of a neuron. In the following it is assumed that

$$g(t) = e^{-t/\tau} . \tag{7.4.1}$$

Let $h(x,t)$ be defined as

$$h(x,t) = \lim_{\Delta,\Delta_1,\Delta_2 \to 0} \text{Prob}\,[X(t) < x < X(t+\Delta) < x+\Delta_1 \,|\, 0 < X(u) < K, \forall u \,\varepsilon\, (0,t), \text{ a firing in } (-\Delta_2, 0)]/\Delta\Delta_1. \qquad (7.4.2)$$

Then h(x,t) satisfies

<u>Lemma 7.4.1</u>

$$h(x,t) = \delta(x-1)\, f(t) + \int_{\max(0,t-\ln(K/(x-1)))}^{t} h(\,(x-1)e^{t-u},\, u\,)\, f(t-u)\, du. \qquad (7.4.3)$$

Using this lemma, one can arrive at

<u>Theorem 7.4.1</u>. The interval pdf P(·) of the firings in Model 7.6 is given by

$$P(t) = \int_{\max(0,t-\ln(K/(K-1)))}^{t} f(t-u)\, du \int_{(K-1)\exp(t-u)}^{K} h(x,u)dx. \qquad (7.4.5)$$

This is obviously a very difficult equation to solve because of the inherent nonlinearity. For a complete discussion of this model, see Srinivasan and Vasudevan (1969).

7.5 Discussion

The main advantage of counter models is the realistic way in which refractoriness can be taken into account through dead time effects. Counter models however have the inherent disadvantage that inhibition cannot be included easily. This makes them conceptually inferior to other types of models. Though it appears that inhibition has been taken into account in Sections 7.2 and 7.3, this is not really so because the models in these sections are only deletion models and therefore have the defects of the latter. In fact counter models are not really neuron models by themselves; they are only the application of studies in counters to other types of models.

CHAPTER 8

DISCRETE STATE MODELS

The interaction of quantal EPSPs and IPSPs at the membrane of the neuron is similar to the birth and death process in probability theory. Thus the membrane potential can occupy only discrete levels increasing by one unit with the arrival of an excitatory impulse and decreasing by one unit with the arrival of an inhibitory impulse. When the potential reaches a level N corresponding to the threshold of firing, the neuron fires and the membrane potential returns to the rest level immediately. The analogy of course is not perfect because the IPSP can take the membrane potential below the rest level. Nevertheless, the theory of birth and death processes can be suitably modified to take this into account. For this reason, in Section 8.1 basic results from the theory of such processes are presented, following Srinivasan and Mehata (1976). Birth and death processes have been studied over many decades and the literature is vast. This section merely introduces the reader to the study of such processes in relation to neuron modelling. It is important to note here that many discrete state models in the literature go far beyond this; in fact it is difficult to identify some of them as birth and death processes. In Sections 8.2 to 8.4 the different types of discrete state models are discussed in detail. At the end of the chapter an assessment of the models is given.

8.1 Birth and death processes

Consider a discrete-valued stochastic process $X(t)$. Let $X(t)$ change state due to two types of events: if $X(t)$ increases by one unit, a *birth* is said to occur; if it decreases by one unit, a *death* is said to occur. The analogy with excitation and inhibition is obvious. Let $X(t)$ satisfy the following properties:

1) $X(t)$ is a stationary time-homogeneous Markov process.

If
$$\pi(n_2,t_2|n_1,t_1) = \text{Prob}\,[X(t_2)=n_2|\,X(t_1)=n_1],\quad t_2>t_1, \tag{8.1.1}$$

2)
$$\begin{aligned}\pi(n,t+\Delta t|m,t) &= \lambda_m(t)\Delta t + o(\Delta t), \quad n=m+1\\ &= \mu_m(t)\Delta t + o(\Delta t), \quad n=m-1\\ &= 1-\big(\lambda_n(t)+\mu_n(t)\big)\Delta t + o(\Delta t), \quad n=m\\ &= o(\Delta t), \quad \text{otherwise}.\end{aligned} \tag{8.1.2}$$

The distribution of $X(t)$ can now be derived using the Chapman - Kolmogorov relation. Thus consider $X(t)$ in the interval $(t,t+\Delta t)$. First it is observed that due to the time-homogeneity of the process,

$$\pi(n_2,t_2|n_1,t_1) = \pi(n_2,t_2-t_1|n_1,0) \tag{8.1.3}$$

$$= \pi(n_2,t_2-t_1|n_1). \tag{8.1.4}$$

One can then write

$$\begin{aligned}\pi(n,t+\Delta t|m) &= [1-(\lambda_n(t)+\mu_n(t))\Delta t]\;\pi(n,t\,|\,m)\\ &+ [\lambda_{n-1}(t)\;\Delta t + o(\Delta t)]\;\pi(n-1,t\,|\,m)\\ &+ [\mu_{n+1}(t)\;\Delta t + o(\Delta t)]\;\pi(n+1,t\,|\,m).\end{aligned} \tag{8.1.5}$$

This equation holds for any state that is not an absorbing state. Consider $n = 0$ an absorbing state. Then (8.1.5) holds for $n \geq 1$. For $n = 0$,

$$\begin{aligned}\pi(0,t+\Delta t|m) &= [1-\lambda_0(t)\;\Delta t]\;\pi(0,t\,|\,m)\\ &+ \mu_1(t)\;\Delta t\;\pi(1,t|m) + o(\Delta t).\end{aligned} \tag{8.1.6}$$

Then, one can obtain from (8.1.5) and (8.1.6)

$$\frac{d\pi}{dt}(n,t \mid m) = -[\lambda_n(t) + \mu_n(t)]\,\pi(n,t \mid m) + \lambda_{n-1}(t)\,\pi(n-1,t \mid m) + \mu_{n+1}(t)\,\pi(n+1,t \mid m),\ n \geqslant 1, \qquad (8.1.7)$$

and

$$\frac{d\pi}{dt}(0,t \mid m) = -\lambda_0(t)\,\pi(0,t \mid m) + \mu_1(t)\,\pi(1,t \mid m). \qquad (8.1.8)$$

The initial condition is

$$\pi(n,0 \mid m) = 1, \quad n = m$$
$$= 0, \quad \text{otherwise.} \qquad (8.1.9)$$

These differential equations describe a <u>non-homogeneous</u> birth and death process because the probability of a birth (death) depends on the state of the process. They may be solved through the use of either the generating function or the Laplace transform.

<u>Theorem 8.1.1</u> For any initial state $m \geqslant 0$ at time $t = 0$ and $|u| < 1$, the generating function $G(t, u)$ of $\pi(n,t \mid m)$ satisfies the partial differential equation

$$G_t(t,u) = \sum_{n=0}^{\infty} u^n\,\pi(n,t \mid m) \left\{(u-1)\,\lambda_n(t) + (u^{-1}-1)\,\mu_n(t)\right\} \qquad (8.1.10)$$

with the initial condition

$$G(0,u) = u^m \qquad (8.1.11)$$

if

$$\text{Prob}\,[X(0) = m] = 1. \qquad (8.1.12)$$

Proof: Multiply (8.1.7) and (8.1.8) by u^n and sum over all values of n to get

$$\frac{\partial}{\partial t} \sum_{n=0}^{\infty} \pi(n,t \mid m)\, u^n = - \sum_{n=0}^{\infty} u^n \pi(n,t \mid m)$$

$$(\lambda_n(t) + \mu_n(t)) + \sum_{n=0}^{\infty} u^n \pi(n-1,t \mid m)\, \lambda_{n-1}(t)$$

$$+ \sum_{n=0}^{\infty} u^n \pi(n+1, t \mid m)\, \mu_{n+1}(t). \tag{8.1.13}$$

After some adjustment

$$G_t(t,u) = - \sum_{n=0}^{\infty} u^n \lambda_n(t)\, \pi(n,t \mid m)$$
$$- \sum_{n=0}^{\infty} u^n \mu_n(t)\, \pi(n,t \mid m) + u \sum_{n=0}^{\infty} u^n \lambda_n(t)\, \pi(n,t \mid m)$$
$$+ u^{-1} \sum_{n=0}^{\infty} u^n \mu_n(t)\, \pi(n,t \mid m), \tag{8.1.14}$$

which, on using 8.1.11 and simplifying, leads to (8.1.10).

Corollary 1 If

$$\lambda_n(t) = n\, \lambda(t) \tag{8.1.15}$$

and

$$\mu_n(t) = n\mu(t), \tag{8.1.16}$$

$$G_t(t,u) = G_u(t,u) \left\{ (u-1)\, [u\, \lambda(t) - \mu(t)] \right\}. \tag{8.1.17}$$

To prove this it is sufficient to observe that

$$\sum_{n=0}^{\infty} n u^n \pi(n,t \mid m) = u\, G_u(t,u), \tag{8.1.18}$$

using which (8.1.17) is obtained.

<u>Theorem 8.1.2</u> If

$$\lambda_n(t) = n\lambda \tag{8.1.19}$$

and

$$\mu_n(t) = n\mu \tag{8.1.20}$$

then

$$G(t,u) = \left\{\frac{\alpha(t) + [1 - \alpha(t) - \beta(t)]u}{1 - \beta(t)\, u}\right\}^m, \tag{8.1.21}$$

where

$$\alpha(t) = \frac{1 - e^{(\lambda - \mu)t}}{\mu - \lambda e^{(\lambda - \mu)t}} \tag{8.1.22}$$

and

$$\beta(t) = \frac{\lambda}{\mu}\,\alpha(t). \tag{8.1.23}$$

This birth and death process is called the <u>linear</u> growth model, the reason for which terminology is obvious from equations (8.1.19) and (8.1.20).

<u>Proof</u>: Using (8.1.19) and (8.1.20) in (8.1.17) one can obtain

$$G_t(t,u) = G_u(t,u)\,(u-1)\,(u\lambda - \mu) \tag{8.1.24}$$

which first order partial differential equation can be solved using the Lagrange method. Thus the Lagrange equations for (8.1.24) are

$$\frac{dt}{1} = \frac{du}{(1-u)\,(\lambda u-\mu)} = \frac{dG}{0} \tag{8.1.25}$$

The general solution is obtained by solving (8.1.25). Thus

$$G(t,u) = \phi\left(\frac{1 - u}{\lambda u - \mu}\; e^{(\lambda - \mu)t}\right) \tag{8.1.26}$$

where $\phi(\cdot)$ is an arbitrary function. Using the initial condition (8.1.11)

$$\phi\left(\frac{1-u}{\lambda u-\mu}\right) = u^m \tag{8.1.27}$$

for all $|u| < 1$.
Thus for all θ such that

$$|1+\theta\mu| < |1+\theta\lambda| ,$$

$$\phi(\theta) = \left(\frac{1+\mu\theta}{1+\lambda\theta}\right)^m . \tag{8.1.28}$$

Replacing θ by $(1-u)\, e^{(\lambda-\mu)t}/(\lambda u-\mu)$ in (8.1.28) and simplifying, equation (8.1.21) is obtained.

The restriction imposed by (8.1.2) can be relaxed by allowing transitions to non-neighbouring levels to obtain a more general type of birth and death process.
Thus let

$$\pi(n,t+\Delta t|i,t) = \rho_{in}(t)\,\Delta t, \tag{8.1.29}$$

from which one can obtain

$$\pi(n,t+\Delta t|m) = \pi(n,t|m)\Big(1-\sum_{i\neq n}\rho_{ni}(t)\,\Delta t\Big)$$

$$+\sum_{i\neq n}\pi(i,t|m)\,\rho_{in}(t)\,\Delta t. \tag{8.1.30}$$

This gives

$$\frac{d}{dt}\pi(n,t|m) + \pi(n,t|m)\sum_{i\neq n}\rho_{ni}(t) = \sum_{i\neq n}\pi(i,t|m)\,\rho_{in}(t) \tag{8.1.31}$$

which can be written as a matrix equation

$$\left[\frac{d\pi_n}{dt}\right] + \left[\pi_n\sum_{i\neq n}\rho_{ni}\right] = \left[\sum_{i\neq n}\pi_i\,\rho_{in}\right] . \tag{8.1.32}$$

With appropriate initial and boundary conditions π_n can be solved for.

It is obvious from the above discussion that by assuming suitable functions for $\lambda(t)$ and $\mu(t)$, a variety of processes can be generated. However it must be clear that all these processes are Markov processes. Hence birth and death processes as studied in the mathematical literature are useful in neuron modelling only when the input sequences are Markov. Nevertheless they have led to fruitful results, though it is necessary to go outside the framework of Markov processes to develop more general neuron models. In the following sections models using input sequences that are Poisson as well as those with non-Poisson sequences are discussed. Many of these models are not easily identifiable as birth and death processes of the kind described earlier in this section. This arises from the need for a variety of models to describe widely varying behaviour in many types of neurons. Nevertheless the basic structure of the birth and death concept remains in most of these models — a birth corresponds to an EPSP, a death to an IPSP. Some artifices have been introduced by various authors to take into account properties like membrane potential decay and the result is a very large number of formulations. Only a representative few will be studied in this chapter and these are divided into three kinds for convenience :

a) models in which there are only excitatory inputs

b) models in which the e-sequence and the i-sequence are independent of each other

c) models in which one sequence is controlled by the other.

8.2 Models with excitatory inputs only

Early stochastic models of neurons considered the cumulative effect of only e-events at the membrane. The neuron model then conveniently reduces to a pure birth process starting at the rest level (zero) and terminating at the threshold N. One of the first modifications of the model was due to Rapoport who considered e-events arriving as a Poisson stream at the neuron membrane resulting in an output spike only if N e-events arrive within an interval of length γ not overlapping with the refractory period following generation of a spike. Ten Hoopen and Reuver (1965a) showed that Rapoport's results are incorrect and modified the model as discussed later in this paragraph. Leslie (1969), in his model of clustering of Poisson points, assumed an output spike to occur with the N-th of N successive e-events provided the interval between any two successive events in the cluster does not exceed a value γ and no e-event in the cluster previous to the N-th is itself a firing. Following Barlow (1963), who suggested that the lifetime of a primary event be considered a r.v. rather than a constant, Ten Hoopen and Reuver (1965a) took a fresh look at Rapoport's model and were led to a set of difference equations. (This concept of the 'lifetime' of an EPSP is discussed in Section 10.1). Lee (1974) starting with the basic deletion model (Model 5.1) built on it by introducing an integrating effect. Some of these models are discussed below. The common features of these models are the following:

1) The neuron receives a sequence of e-events, each of which gives rise to an EPSP of unit size.
2) The e-events form a renewal process with interval pdf $f(\cdot)$. (In the models in this section the e-events are assumed to constitute a Poisson process with parameter λ).
3) When the membrane potential $X(t)$ reaches a level N, the neuron fires and the membrane potential immediately returns to the rest level (assumed zero for convenience).

For this basic model the first passage time pdf is given by

<u>Theorem 8.2.1</u>. The firings in the basic discrete state neuron model form a renewal process with pdf $P(\cdot)$ given by

$$P(\cdot) = f_N(\cdot), \tag{8.2.1}$$

where the subscript stands for convolution of order N.

<u>Corollary</u> If

$$f(t) = \lambda e^{-\lambda t}, \tag{8.2.2}$$

then

$$P(t) = \frac{\lambda(\lambda t)^{N-1} e^{-\lambda t}}{(N-1)!}, \tag{8.2.3}$$

which is a gamma distribution of order N.

8.2.1 Model 8.1 (Rapoport, 1950; Ten Hoopen and Reuver, 1965a)

Rapoport made the additional assumption

4) a firing occurs only when N e-events occur within an interval of length γ.

Thus if there occurs an e-event in $(t,t+dt)$ and the neuron has not fired in $(t,t+\gamma)$ then $X(t)$ decreases by one unit in $(t,t+\gamma+dt)$. That is, the post-synaptic effect of an e-event has a 'lifetime' γ. Ten Hoopen and Reuver (1965a) showed that Rapoport's analysis is defective and took a different approach. They assumed that the 'lifetime' is a r.v. uniformly distributed over $(0,\gamma)$. Thus the probability of an e-event 'dying' at time $t > \gamma$ if there are k e-events 'alive' is simply $k dt/\gamma$.
Let

$$\pi_k(t) = \text{Prob [k e-events are 'alive' at t]}, \quad k < N. \tag{8.2.4}$$

Then

$$\pi_k(t) = \frac{\lambda(\lambda t)^{k-1} e^{-\lambda t}}{(k-1)!}, \quad 0 \leqslant k < N, \ 0 \leqslant t \leqslant \gamma. \tag{8.2.5}$$

For $t > \gamma$

$$\frac{d\pi_0(t)}{dt} = -\lambda \pi_0(t) + \frac{1}{\gamma} \pi_1(t), \tag{8.2.6}$$

$$\frac{d\pi_k(t)}{dt} = \lambda \pi_{k-1}(t) - (\lambda + \frac{k}{\gamma}) \pi_k(t) + \frac{k+1}{\gamma} \pi_{k+1}(t), \quad 0 < k < N-1, \tag{8.2.7}$$

and

$$\frac{d\pi_{N-1}(t)}{dt} = \lambda \pi_{N-2}(t) - (\lambda + \frac{N-1}{\gamma}) \pi_{N-1}(t), \tag{8.2.8}$$

with initial conditions

$$\pi_0(\gamma) = e^{-\lambda\gamma}, \tag{8.2.9}$$

$$\pi_k(\gamma) = \frac{(\lambda\gamma)^k e^{-\lambda\gamma}}{k!}, \quad 0 < k \leqslant N-1. \tag{8.2.10}$$

These results are easily obtained by making use of the fact that the probability of an e-event occurring in $(t, t+\Delta t)$ is $\lambda \Delta t$ and the probability of an e-event 'dying' in an interval $(t, t+\Delta t)$, $t > \gamma$, if k e-events are 'alive' at t, is $k\Delta t/\gamma$. The following theorem can be proved.

Theorem 8.2.2 The Laplace transform of the p.d.f. of the first passage time to threshold N (i.e., the firing interval) P(t) in Model 8.1 is given by

$$P^*(s) = [\lambda C_N(s)/D_N(s)]\, e^{-\gamma s} + \left(\frac{\lambda}{\lambda+s}\right)^N (1 - e^{-\gamma s}) \tag{8.2.11}$$

where $D_N(s)$ is given by

$$\begin{aligned} D_k(s) &= 1, \quad k = 0 \\ &= -s-\lambda, \quad k = 1 \\ &= -\left(s+\lambda+\frac{k-1}{\gamma}\right) D_{k-1}(s) \\ &\quad - \frac{\lambda(k-1)}{\gamma} D_{k-2}(s), \quad 1 < k \leqslant N \end{aligned} \tag{8.2.12}$$

and

$$C_N(s) = (-1)^N \lambda^{N-1} e^{-\lambda\gamma} \sum_{k=0}^{N-1} (-\gamma)^k D_k(s)/k! \tag{8.2.13}$$

Using Theorem 8.2.2, one can evaluate the mean firing interval $\bar{T}$ from

$$\bar{T} = \int_0^{\gamma} t\, P(t)\, dt + \int_{\gamma}^{\infty} t\, P(t)\, dt. \tag{8.2.14}$$

Theorem 8.2.3 The mean firing interval in Model 8.1 is given by

$$\bar{T} = \frac{N}{\lambda}\left[1 - e^{-\lambda\gamma} \sum_{k=0}^{N} \frac{(\lambda\gamma)^k}{k!}\right]$$

$$+ \gamma e^{-\lambda\gamma} \sum_{k=0}^{N-1} \frac{(\lambda\gamma)^k}{k!}$$

$$+ \frac{e^{-\lambda\gamma}}{\lambda}\left[1 + \sum_{k=1}^{N-1} \frac{k!}{(\lambda\gamma)^k} \sum_{m=0}^{k} \frac{(\lambda\gamma)^m}{m!}\right.$$

$$\left. + \sum_{k=1}^{N-1} \frac{(\lambda\gamma)^k}{k!} \sum_{m=k}^{N-1} \frac{m!}{(\lambda\gamma)^m} \sum_{n=0}^{m} \frac{(\lambda\gamma)^n}{n!}\right]. \quad (8.2.15)$$

8.2.2 Model 8.2 (Leslie, 1969)

Leslie made the following additional assumption:

4) a firing occurs with the N-th e-event of a group of N e-events if no time gap between successive e-events in this group exceeds γ and no e-event in this group itself gives rise to a firing.

Thus in this model the 'lifetime' of an e-event has a minimum length γ and a maximum length $(N-1)\gamma$. To find the pdf of the interval $P(t)$ between two successive firings, a series of lemmas will now be stated.

Lemma 8.2.1

$$A(t)\,dt = \text{Prob}\,[\text{interval between two e-events is } t, \text{ and } t < \gamma]$$

$$= \lambda e^{-\lambda t} H(\gamma - t), \quad (8.2.16)$$

where $H(\cdot)$ is the Heaviside function,

and $$A^*(s) = \Lambda E \quad (8.2.17)$$

with

$$\Lambda = \lambda/(\lambda + s) \tag{8.2.18}$$

and

$$E = 1 - e^{-(\lambda+s)\gamma}. \tag{8.2.19}$$

<u>Lemma 8.2.2</u> If B(t) = Prob [interval between two successive e-events = γ], then

$$B(t) = \lambda e^{-\lambda t} \quad \delta(t-\gamma), \tag{8.2.20}$$

where $\delta(\cdot)$ is the Dirac delta function, and

$$B^*(s) = \lambda(1 - E). \tag{8.2.21}$$

Now a firing can occur in two mutually exclusive and exhaustive ways:

1) a firing occurs with the N-th e-event after the origin in $(t,t+dt)$, $t \leq (N+1)\gamma$, the first e-event occurring in (t_1,t_1+dt_1), $t_1 < \gamma$. This is labelled 'event A'.
2) a firing occurs in $(t,t+dt)$ with the N-th of a cluster of N e-events such that no firing occurs in $(0,t_1)$, no e-event occurs in $(t_1,t_1+\gamma)$, and the first e-event of the cluster occurs in $(t_1+\gamma,t_1+\gamma+dt_1)$. This is 'event B'.

Then the following lemma can be easily proved using lemmas (8.2.2) and (8.2.3):

<u>Lemma 8.2.3</u>

$$L\,[\text{Prob}\,(A)] = (\Lambda E)^N, \tag{8.2.22}$$

and

L [Prob (B)] = L[Prob (time to a firing < t)] L[Prob (time between two successive e-events = $\tilde{\gamma}$)]

L [Prob (N e-events occur, in a time t, no two events being separated by an interval > γ]

$$= \frac{1 - P^*(s)}{s} \lambda (1-E) (\Lambda E)^{N-1} \qquad (8.2.23)$$

where L stands for the Laplace transform.

The Laplace transform of the p.d.f. of the firing interval is now given by

Theorem 8.2.4. The Laplace transform of the pdf of the interval between two successive firings in Model 8.2 is

$$P^*(s) = \frac{\Lambda (\Lambda E)^{N-1}(1 - \Lambda E)}{(1-\Lambda E) - \Lambda(1-E)(1 - (\Lambda E)^{N-1})} . \qquad (8.2.24)$$

Equation (8.2.24) is obtained by simply adding (8.2.22) and (8.2.23) and simplifying.

Corollary: The mean and the variance of the firing interval T are given by

$$\bar{T} = E(T) = \frac{1}{\lambda} \frac{1 - (1 - \xi)^N}{\xi (1 - \xi)^{N-1}} \qquad (8.2.25)$$

and

$$\text{Var}(T) = \left(\frac{1}{\xi \lambda (1 - \xi)^{N-1}}\right)^2$$

$$[(\xi^2 + 2\lambda\gamma\xi - 1)(1 - \xi)^{2N-2}$$

$$- 2(N-1)\xi(1 - \xi)^{N-1}$$

$$+ 2\lambda\gamma\xi(1 - \xi)^{N-2}(N\xi - 1) + 1] \qquad (8.2.26)$$

respectively, where

$$\xi = e^{-\lambda\gamma}. \tag{8.2.27}$$

Leslie has inverted the Laplace transform in (8.2.24) and has given extensive numerical results for different values of λ, γ and N.

8.3 Models with independent interaction of e-events and i-events

Since inhibition is an important part of signal processing in the nervous system, it has to be included in any realistic neuron model. Several models of neurons in which sequences of e- and i-events interact independently have been proposed. Fetz and Gerstein (1963) and Stein (1965) introduced inhibitory impulses arriving as a Poisson process into the pure birth process model assuming that the membrane potential decays exponentially with a time constant τ. However this decay effect makes the mathematics extremely cumbersome. Hence these studies were confined mostly to simulation with a noise generator or on a computer. Following this, Ten Hoopen (1966b) removed the deterministic decay effect and substituted for it a 'probabilistic decay'. In this approach, the membrane potential is assumed to move one unit towards the rest level in the absence of input impulses with a probability $\frac{n}{\tau}\Delta t$ in an interval $(t, t + \Delta t)$. Effectively this introduces an inhomogeneous Poisson process into the model. This aspect will be discussed later. An analytic solution of the first passage time problem was given only for the cases of infinite threshold and no decay. Following this, Goel, Richter-Dyn and Clay (1972) obtained analytical expressions for the mean and the variance of the firing interval in Ten Hoopen's model and extended the model by postulating that the the probability of the membrane potential reaching the threshold N when it is above the rest level increases as it approaches N.

In an entirely different approach, Hochman and Fienberg (1971) modified Model 8.2 (Leslie, 1969) by introducing different mechanisms of inhibition. The main significance of these models is their propensity to generate multimodal distributions (Fienberg and Hochman, 1972).

Ten Hoopen and Reuver (1967a) have considered a different kind of inhibitory mechanism. In crustacean muscle fibers and stretch receptors, inhibitory impulses on special inhibitory nerves have the effect of clamping the membrane potential to the rest level (Ochs, 1965, p. 182). In Ten Hoopen and Reuver's model, e-events increase the membrane potential by one unit and if the potential reaches the threshold N the neuron fires. There is a sequence of i-events independent of the e-events, each i-event causing the potential to be reset to zero, so that integration of EPSPs must start anew with the next e-event. Such a reset mechanism conveniently eliminates negative potentials and leads to a random walk model with two finite boundaries — the rest level and the threshold. These models are discussed below.

8.3.1 Model 8.3 (Fetz and Gerstein, 1963; Stein, 1965; Ten Hoopen, 1966b; Goel, Richter-Dyn and Clay, 1972).

The introduction of inhibition into the pure birth process does not lead strictly to a birth and death process because inhibition can cause the membrane potential to go below the rest level, i.e. it can hyperpolarise the membrane. Hence the analysis in Section 8.1 should be modified slightly.

First, the model is assumed to have the following properties:

1) The neuron receives a Poisson stream of e-events with parameter λ and Poisson i-events with parameter μ.
2) $X(t)$, the membrane potential, increases one step with an e-event and decreases one step with an i-event.
3) In the absence of inputs, $X(t)$ decreases one step in time $(t, t+\Delta t)$ with probability $n\,\Delta t/\tau + o(\Delta t)$ when $X(t) = n < 0$, and increases one step with probability $-\,n\,\Delta t/\tau + o(\Delta t)$ when $X(t) = n < 0$.
4) When $X(t)$ reaches $N+1$, the neuron fires and is reset to the rest level zero.

With the above assumptions, (8.1.29) can be written as

$$\begin{aligned} \pi(n, t + \Delta t \mid m, t) &= \rho_{mn}\,\Delta t + o(\Delta t), \quad n = m+1,\ m-1 \\ &= 1 - \rho_{mn}\,\Delta t + o(\Delta t), \quad n = m \\ &= o(\Delta t), \quad \text{otherwise} \end{aligned} \tag{8.3.1}$$

so that the membrane potential can now be described by

Lemma 8.3.1.

$$\frac{d\pi_n(t)}{dt} = \rho_{n+1,n}\ \pi_{n+1}(t) - (\rho_{n,n+1} + \rho_{n,n-1})\ \pi_n(t)$$

$$+ \rho_{n-1,n}\ \pi_{n-1}(t), \quad n \leqslant N - 1 \tag{8.3.2}$$

and

$$\frac{d\pi_N(t)}{dt} = -(\rho_{N,N+1} + \rho_{N,N-1})\ \pi_N(t) + \rho_{N-1,N}\ \pi_{N-1}(t), \tag{8.3.3}$$

where $\pi(n,t|m)$ is written $\pi_n(t)$ assuming that the initial state m is the rest level equal to zero. Equations (8.3.2) and (8.3.3) can now be written compactly in matrix notation

$$\frac{d\tilde{\pi}}{dt}(t) = \tilde{\rho}\,\tilde{\pi}(t) \tag{8.3.4}$$

where

$$\tilde{\pi} = [\ldots,\ \pi_{-1},\ \pi_0,\ \pi_1, \ldots, \pi_N], \tag{8.3.5}$$

and

$$\tilde{\rho} = \begin{bmatrix} \ddots & & & & & \\ \rho_{-1,0} & -(\rho_{0,1}+\rho_{0,-1}) & \rho_{1,0} & & & \\ & \rho_{0,1} & -(\rho_{1,2}+\rho_{1,0}) & \rho_{2,1} & & \\ & & \ddots & \ddots & \ddots & \\ & & & \ddots & \ddots & \rho_{N,N-1} \\ & & & & \rho_{N-1,N} & -(\rho_{N,N+1}+\rho_{N,N-1}) \end{bmatrix} \tag{8.3.6}$$

which is a matrix of infinite size because there is no barrier below the rest level.

<u>Theorem 8.3.1</u> The formal solution of the state equation describing the membrane potential is

$$\tilde{\pi}(t) = e^{\rho t}\,\tilde{\pi}(0), \tag{8.3.7}$$

the initial state vector $\tilde{\pi}(0)$ being given by

$$\begin{aligned} \pi_n(0) &= 1 \;, \quad n = 0 \\ &= 0 \;, \quad \text{otherwise} . \end{aligned} \tag{8.3.8}$$

<u>Theorem 8.3.2</u> The interval pdf of the firing sequence in Model 8.3 is given by

$$P(t) = -\frac{d}{dt} \sum_{-\infty}^{N} \pi_n(t). \tag{8.3.9}$$

This result is obtained by adding (8.3.2) and (8.3.3) and using the fact

$$P(t) = \rho_{N,N+1}\, \pi_N(t). \tag{8.3.10}$$

More explicit results are given below. Since the analysis is lengthy the results are stated without proof. For a complete analysis of the model see Goel et al. (1972).

The transition probabilities are first written in the proper form. This is given by

$$\begin{aligned} \rho_{n,n+1} &= \lambda \,, \quad 0 \leqslant n \leqslant N \\ &= \lambda - n/\tau \,, \quad n < 0 \end{aligned} \tag{8.3.11}$$

and

$$\rho_{n,n-1} = \mu + n/\tau \ , \quad 0 \leqslant n \leqslant N$$

$$= \mu \quad , \quad n < 0 \ . \tag{8.3.12}$$

For these transition probabilities, analytic expressions for $\pi_n(t)$ cannot be obtained, except when $\tau \to \infty$, which is the case of no 'decay'. For this special case, $\pi_n(t)$ is given by

Theorem 8.3.3. The pdf of the first passage time in Model 8.3 is given by

$$P(t) = (N+1)(\lambda/\mu)^{(N+1)/2} t^{-1} e^{-(\lambda+\mu)t}$$

$$I_{N+1}(2\sqrt{\lambda\mu}\, t) , \tag{8.3.13}$$

where $I,(\cdot)$ is the Bessel function of order N+1.

Corollary The mean and the variance of the first passage time are given by

$$\bar{T} = (N+1)/(\lambda-\mu) \tag{8.3.14}$$

and

$$\mathrm{Var}\ (T) = (N+1)(\lambda+\mu)/(\lambda-\mu)^3 \ . \tag{8.3.15}$$

When $\tau \neq \infty$, though the pdf of the first passage time cannot be obtained explicitly, the mean and the variance can. Theorem 8.3.4 states these results.

Theorem 8.3.4. In Model 8.3, the mean and the mean square of the first passage time are given by

$$\bar{T}/\tau = \frac{\Phi(1,a+1,b)-1}{a\,\Gamma(b+1)} \sum_{j=0}^{N} \frac{\Gamma(b+j+1)}{a^j}$$

$$+ \Gamma(b+N+2) \sum_{j=1}^{N+1} \frac{a^{-j}}{j\ \Gamma(b+N+i-j)}$$

$$- \frac{b}{\Gamma(b+1)} \sum_{j=1}^{N+1} \frac{a^{-j}}{j} \Gamma(b+j+1), \tag{8.3.16}$$

and

$$\bar{T}^2/2 = \sum_{j=0}^{\infty} s(N-j)\, v(N-j), \tag{8.3.17}$$

where

$$\begin{aligned} a &= \lambda \tau, \\ b &= \mu\tau , \end{aligned} \tag{8.3.18}$$

the $s(j)$ satisfy the difference equation

$$s(j) = \frac{\rho_{j,j-1}}{\rho_{j,j+1}} s(j-1) + \frac{1}{\rho_{j,j+1}}, \quad j \leq N \tag{8.3.19}$$

$v(i)$ satisfies

$$v(i-1) = [(\rho_{i,i-1} + \rho_{i+1,i})\, v(i) - \rho_{i+1,i}\, v(i+1)]/\rho_{i,i-1} + K_i , \tag{8.3.20}$$

with

$$\begin{aligned} K_i &= 1/\rho_{i,i-1} \quad , \quad i \geq 0 \\ &= 0 \qquad\qquad , \quad i < 0 \end{aligned} \tag{8.3.21}$$

and Φ is the hypergeometric function.

Model 8.3 extended

Goel, Richter-Dyn and Clay (1972) extended the above model by making the additional assumption:

5) As the membrane potential nears the threshold, the probability of reaching the threshold increases. Thus it is assumed that the probability of an e-event occurring in $(t,t+\Delta t)$ is $n\lambda\Delta t + o(\Delta t)$.

With this, equations (8.3.11) and (8.3.12) become

$$\rho_{n,n+1} = (n+1)\lambda \;, \quad 0 \leq n \leq N$$
$$= \lambda - n/\tau \;, \quad n < 0 \tag{8.3.22}$$

$$\rho_{n,n-1} = \mu + n/\tau \;, \quad 0 \leq n \leq N$$
$$= \mu \;, \quad n \leq 0 . \tag{8.3.23}$$

In this model too, the first passage time pdf cannot be obtained analytically. Unfortunately even the mean and the mean square can not be obtained easily. However, if inhibition is absent the model is simplified and for this the interval pdf of the intervals can be obtained.

<u>Theorem 8.3.5</u> In the extended Model 8.3 if inhibition is absent, i.e., if $\mu = 0$, the pdf of the interval between successive firings is given by

$$P(t) = \frac{(N+1)\, a^{1-\nu}}{\tau} \sum_{j=0}^{N} d_j \exp\left\{ -(1-a)\, \mu_j\, t/\tau \right\} \tag{8.3.24}$$

where

$$\nu = \text{rest level} \tag{8.2.25}$$

$$d_j = \ell_\nu(\mu_j)\, \ell_N(\mu_j) / \sum_{s=0}^{N} \ell_s^2(\mu_j)\, a^{-s} , \tag{8.3.26}$$

and $\ell_n(\mu)$ are the Gottlieb polynomials with the μ_js being the roots of the equation

$$\ell_{N+1}(\mu) = 0 . \tag{8.3.27}$$

<u>Corollary</u> The mean and the mean square of the firing interval are given by

$$\bar{T} = \frac{(N+1)\, a^{1-\nu}}{\tau} \sum_{j=0}^{N} d_j \left\{ (1-a)\, \mu_j/\tau \right\}^{-2} , \tag{8.3.28}$$

and

$$\overline{T^2}/2 = \frac{(N+1)\, a^{1-\nu}}{\tau} \sum_{j=0}^{N} d_j \left\{(1-a)\, \mu_j/\tau\right\}^{-3} . \qquad (8.3.29)$$

Goel, Richter-Dyn and Clay (1972) have given extensive numerical results in the form of graphs relating the input and the output. In addition to these graphs, curves of the coefficient of variation, $(\overline{T^2} - \bar{T}^2)^{\frac{1}{2}}/\bar{T}$ versus $\bar{T}$, as well as the interval pdf in the extended model, are given. For a complete discussion see Goel et al. (1972).

8.3.2 Model 8.4 (Hochman and Fienberg, 1971)

The models in this section are all extensions of Model 8.2 (Leslie, 1969) in which different modes of inhibition are introduced. To start with, consider Model 8.2 in which

1) e-events form a Poisson process with parameter λ
2) a firing occurs with the N-th event of a group of N e-events provided no time gap between successive e-events in this group exceeds γ and no e-event in this group itself gives rise to a firing.

Under these conditions, the interval pdf $Q(\cdot)$ is given by Theorem 8.2.3. For convenience equation (8.2.24) is rewritten here with a slight change in the notation

$$Q^*(s) = \frac{\Lambda\,(\Lambda E)^{N-1}\,(1-\Lambda E)}{1 - \Lambda E - \Lambda(1-E)(1-(\Lambda E)^{N-1})}$$

with

$$\Lambda = \lambda/\lambda + s \qquad (8.3.30)$$

and

$$E = 1 - e^{-(\lambda + s)\gamma} .$$

To this scheme, an inhibitory process is added.

3) The i-events are Poisson with parameter μ.

Model 8.4(1) In this model the following additional assumption is made.

4) A firing occurs due to a cluster N only if no i-event occurs during this cluster.

The firing interval pdf is then given by

Theorem 8.3.6. In Model 8.4(1), the Laplace transform of the firing interval pdf $P_1(\cdot)$ is given by

$$P_1^*(s) = \frac{(s+\mu)\, Q^*(s+\mu)}{s + \mu\, Q^*(s+\mu)}. \tag{8.3.31}$$

Proof: The event of the first firing in (t,t+dt) is the union of two disjoint events, A and B: A — that of a firing in (t,t+dt) and no i-events in (0,t), B — that of a firing in (t,t+dt) and at least one i-event in (0,t).
Then

$$\text{Prob (A)} = Q(t)\, e^{-\mu t}\, dt \tag{8.3.32}$$

and

$$\text{Prob (B)} = [\int_0^t e^{-\mu x}\left\{\int_x^\infty Q(u)\, du\right\} \mu P_1(t-x)\, dx]\, dt \tag{8.3.33}$$

so that

$$P_1(t)\, dt = \text{Prob (A)} + \text{prob (B)}. \tag{8.3.34}$$

From this, $P_1^*(s)$ is found to be given by

$$P_1^*(s) = Q^*(s+\mu) + \mu\, P_1^*(s)\, \frac{[1-Q^*(s+\mu)]}{s+\mu}. \tag{8.3.35}$$

Simplifying, (8.3.31) is obtained.

Corollary The mean and the variance of the firing interval are given by

$$\bar{T}_1 = \frac{(\Lambda_\mu)^{-N} - (E_\mu)^N}{(E_\mu)^{N-1}\,[\mu + \lambda(1 - E_\mu)]} \tag{8.3.36}$$

and

$$\text{Var}\,(T_1) = \frac{1 + 2\mu\,\dot{Q}^*(\mu) - [Q^*(\mu)]^2}{[\mu Q^*(\mu)]^2} \tag{8.3.37}$$

where the dot represents differentiation w.r.t. s and Λ_μ and E_μ are given by (8.3.30) with $s = \mu$.

Further modifications in the model are made below. Since the proofs are all fairly similar to those in Model 8.4 (1), only the results will be stated.

Model 8.4(2) This is the same as Model 8.4 (1) with the following feature added —

5) Every i-event is followed by a dead time during which neither e-events nor i-events are registered. The dead time is a r.v. with pdf $\eta_2(\cdot)$.

Theorem 8.3.7. The interval pdf $P_2(\cdot)$ in Model 8.4 (2) is given by its Laplace transform

$$P_2^*(s) = \frac{(s+\mu)\,Q^*(s+\mu)}{s+\mu\,[1-\eta_2^*(s)] + \mu\eta_2^*(s)\,Q^*(s+\mu)}\,. \tag{8.3.38}$$

Corollary. If $\bar{T}_d$ is the mean of the dead time, then the mean and the variance of the firing interval are given by

$$\bar{T}_2 = \bar{T}_1\,(1 + \mu\,\bar{T}_d) \tag{8.3.39}$$

and

$$\text{Var}\,(T_2) = \text{Var}\,(T_1)\,(1 + \mu\,\bar{T}_d)$$

$$+ \mu\,[\bar{T}_1\,\text{Var}\,(T_d) + \bar{T}_d\,\bar{T}_1^{\,2}$$

$$+ \bar{T}_d^{\,2}\ \bar{T}_1] + \mu^2\ \bar{T}_1^2\ \bar{T}_d^2 \ . \qquad (8.3.40)$$

<u>Model 8.4(3)</u> This is the same as Model 8.4 (1) but in addition has the property

5) An i-event gives rise to a dead-time that is a r.v. with pdf $\eta_3(\cdot)$ whether or not it arrives during a dead-time.

<u>Lemma 8.3.3</u> The effective dead time following an i-event that is registered is a random variable that has a pdf $\theta(\cdot)$ with Laplace transform given by

$$\theta^*(s) = \frac{(s + \mu)\,\eta_3^*(s + \mu)}{s + \mu\,\eta_3^*(s + \mu)} \ . \qquad (8.3.41)$$

<u>Theorem 8.3.8</u> The interval pdf $P_3(\cdot)$ in Model 8.4 (3) is given in terms of its Laplace transform as

$$P_3^*(s) = \frac{[s + \mu\,\eta_3^*(s + \mu)]\ Q^*(s+\mu)}{s + \mu\,\eta_3^*(s + \mu)\ Q^*(s+\mu)} \ . \qquad (8.3.41)$$

<u>Corollary</u> The mean and the variance of the firing interval are given by

$$\bar{T}_3 = \bar{T}_1/\eta_3^*(\mu), \qquad (8.3.42)$$

and

$$\mathrm{Var}\,(T_3) = \dot{\eta}_3^*(\mu).\ \mathrm{Var}\,(T_1)$$

$$+ \frac{[1 - \eta_3^*(\mu)]\bar{T}_1^2}{[\eta_3^*(\mu)]^2}$$

$$+ \frac{1 - \eta_3^*(\mu) + \mu\dot{\eta}_3^*(\mu)}{\mu\,[\eta_3^*(\mu)]^2}\,(2\ \bar{T}_1) \ . \qquad (8.3.43)$$

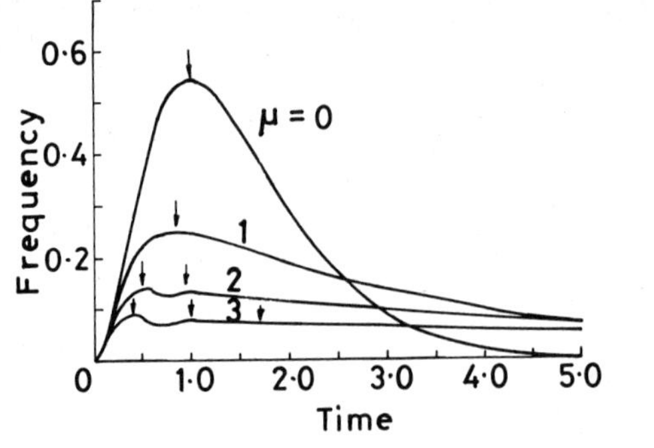

(a) $\lambda = 2,\ N = 3,\ \gamma = 3,\ \eta_2(t) = \delta(t - 0{\cdot}5)$

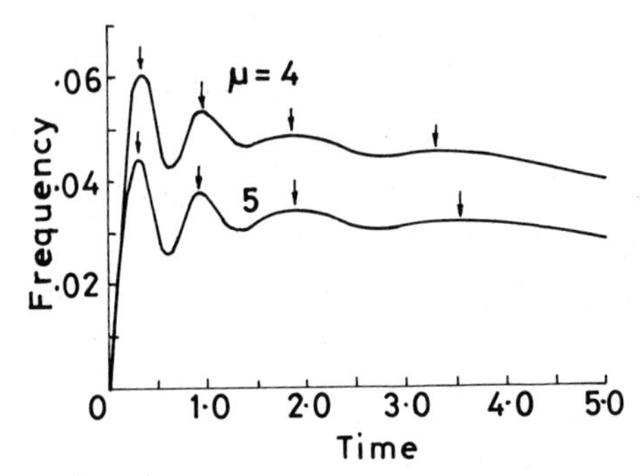

(b) Same as in (a) with vertical axis expanded

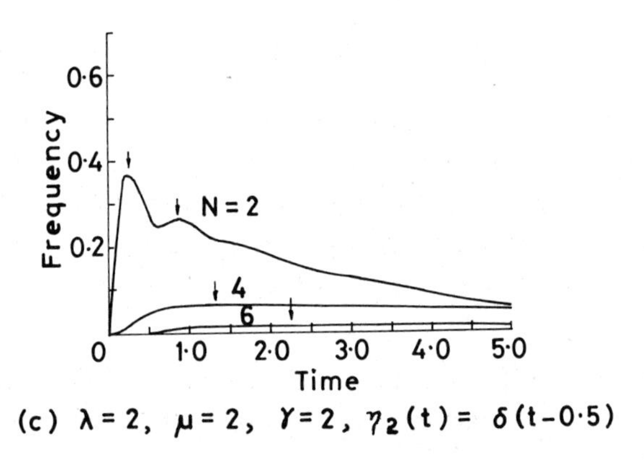

(c) $\lambda = 2,\ \mu = 2,\ \gamma = 2,\ \eta_2(t) = \delta(t - 0{\cdot}5)$

FIG. 8.3.1. MULTIMODAL DISTRIBUTIONS IN MODEL 8·4 (Fienberg and Hochman, 1972)

In the above, taking a specific form of $\eta_3(\cdot)$,

$$\eta_3(t) = \frac{n}{c}\left(\frac{nt}{c}\right)^n \frac{1}{n!} e^{-nt/c}, \tag{8.3.44}$$

then

$$\frac{1}{\eta_3^*(\mu)} = \left(1 + \frac{\mu c}{n}\right)^{n+1} \tag{8.3.45}$$

and

$$\bar{T}_3 = \bar{T}_1 \left(1 + \frac{\mu c}{n}\right)^{n+1} > \bar{T}_1 \left(1 + \frac{n+1}{n} \mu c\right) = \bar{T}_2. \tag{8.3.46}$$

Of all these models, Model 8.4(2) is the most significant because of its propensity to generate multimodal distributions. Fienberg and Hochman (1972) have shown that if the dead time is a constant (=d) then $P_2(t)$ has modes occurring at

$$t = t_z \tag{8.3.47}$$

and

$$t = t_z + nd + (n-1)/\mu, \; n = 1,2,\cdots, \tag{8.3.48}$$

where t_z is the value at which the single mode of $e^{-\mu t} Q(t)$ occurs. The pdf of $P_2(\cdot)$ is shown in Figure 8.3.1 for different values of N and μ and a constant dead time of length 0.5. Observe that as the inhibition rate increases the modes appear, but as the threshold increases, the modes disappear.

8.3.3 Model 8.5 (Ten Hoopen and Reuver, 1967a; Srinivasan, Rajamannar and Rangan, 1971)

In this model, the inhibitory mechanism is totally different from that considered till now. An i-event does not lead to an IPSP but resets the membrane potential to the reset level. The following are the features of the model.

1) The e-events form a renewal process with pdf $f(\cdot)$, an e-event increasing the membrane potential $X(t)$ by one unit, i.e., the e-event results in an EPSP of unit size.
2) A firing occurs when $X(t)$ reaches N, the threshold value, after which the potential is reset to the rest level, assumed zero.
3) The i-events form a renewal process with pdf $g(\cdot)$. An i-event resets $X(t)$ to zero so that the accumulation has to start anew. Thus a firing can occur only with a run of N e-events uninterrupted by an i-event.

As has been mentioned at the beginning of this section, such a model can be used in neurons in which the inhibitory impulse clamps the potential to the rest level.

It is clear from the formulation that the firing sequence is not a renewal process. Hence the interval pdf does not give complete information about the process. Higher order properties have been derived by Srinivasan, Rajamannar and Rangan (1971). Here only the interval pdf is derived.

<u>Lemma 8.3.4</u> If

$$\pi(t) = \lim_{\Delta, \Delta' \to 0} \text{Prob}[\text{the first i-event after the origin occurs in } (t, t+\Delta') \mid \text{a firing occurs in } (-\Delta, 0)]/\Delta' , \tag{8.3.49}$$

$$\pi(t) = C \int_t^\infty g(w)\, dw \sum_{k=1}^\infty \int_0^{w-t} f_{Nk-1}(v)\, dv \int_{w-t-v}^\infty f(u)\, du \tag{8.3.50}$$

here C is a normalising constant obtained from

$$\int_0^\infty \pi(t)\, dt = 1 \tag{8.3.51}$$

and the subscript to $f(\cdot)$ represents convolution of that order.

<u>Lemma 8.3.5</u> If

$$\pi_k(t,\tau) = \lim_{\Delta,\Delta',\Delta''\to 0} \text{Prob}[\text{ the k-th i-event after the origin occurs in } (\tau,\tau+\Delta'), \text{ the first e-event after this k-th i-event occurs in } (t,t+\Delta'') \text{ and no firing occurs in } (0,t) \mid \text{a firing occurs in } (-\Delta,0)]/\Delta'\Delta'' , \quad (8.3.52)$$

then

$$\pi_1(t,\tau) = \pi(\tau) \int_0^{\tau} f(t-v) \sum_{j=1}^{N-1} f_j(v)\, dv + f(t)\, \pi(\tau), \quad (8.3.53)$$

the subscript j standing for convolution of that order,

$$\pi_2(t,\tau) = \int_0^{\tau} g(\tau-w)\, \pi_1(t,w)\, dw$$

$$+ \iint_{\tau>v>w>0} f(t-v)\, g(\tau-w)\, \pi_1(v,w)\, dw\, dv, \quad (8.3.54)$$

and, for $k > 2$,

$$\pi_k(t,\tau) = \int_0^{\tau} g(\tau-w)\, \pi_{k-1}(t,w)\, dw + \iint_{\tau>v>w>0} f(t-v)\, g(\tau-w)$$

$$\pi_{k-1}(v,w)\, dw\, dv + \iiint_{\tau>v>u>w>0} f(t-v)\, g(\tau-w)$$

$$\sum_{j=1}^{N-2} f_j(v-u)\, \pi_{k-1}(u,w)\, dw\, du\, dv. \quad (8.3.55)$$

<u>Lemma 8.3.6</u> If

$$p_k(t) = \lim_{\Delta,\Delta'\to 0} \text{Prob}[\text{a firing occurs in } (t,t+\Delta') \text{ and k i-events occur in } (0,t) \mid \text{a firing occurs in } (-\Delta,0)]/\Delta', \quad (8.3.56)$$

then

$$p_0(t) = f_N(t) \int_t^{\infty} \pi(v)\, dv,$$

and, for $k \geqslant 1$,

$$p_k(t) = \iint\limits_{t>w>v>0} f_{N-1}(t-w)\, \pi_k(w,v)\, dv\, dw \int_{t-v}^{\infty} g(u)\, du. \tag{8.3.57}$$

Using these lemmas the firing interval pdf can be found.

Theorem 8.3.9. In Model 8.5, the interval between two successive firings has the pdf

$$P(t) = \sum_{k=0}^{\infty} p_k(t). \tag{8.3.58}$$

In this model, if $g(\cdot)$ is exponential the firing sequence is a renewal process and hence is completely described by $P(\cdot)$. If $g(\cdot)$ is not exponential, higher-order properties must be known. Thus the product densities of degree one and two and the pdf governing two successive intervals separated by firings have been derived by Srinivasan et al. (1971). Since the method of approach is similar to that used in the derivation of higher order properties in Model 5.1 (see Section 5.1.2), these results are not presented here. Srinivasan et al. (1971) have also considered extensions of the model by introducing a random lifetime of an i-event. In the first of these models both the e-events and the i-events are Poisson and the lifetimes of i-events are iid r.v.s. The pdf of the interval between two successive firings can be then easily derived. In the second model, the i-events are taken to form a renewal process. Once again the interval pdf of the firing sequence can be derived, though now the firings do not form a renewal process. For the details see Srinivasan et al.(1971).

8.4 Models with dependent interaction of input sequences

In all the models discussed before, the excitatory and the inhibitory sequences are assumed to be independent of each other.

Strictly speaking, this is not valid because neurons are interconnected. Ten Hoopen and Reuver (1967a) introduced a kind of dependent interaction into one of their models in an attempt to model certain types of neurons in which the output of one neuron triggers a sequence of firings in another. Such behaviour occurs in the spinal cord in which inhibition occurs by local feedback through Renshaw cells (see p. 62). Ten Hoopen and Reuver called their model a 'delay model' which is almost identical to Model 8.5, which they call 'deletion model'. This terminology is unfortunately confusing because the <u>deletion models</u> of Chapter 5 are in a class by themselves because they model a specific form of inhibition which is termed deletion. In Model 8.5, the effect of an i-event is to reset $X(t)$ to zero rather than to delete an e-event. Further, in the deletion models of Chapter 5, deletion or pre-inhibition is of an e-event and is effected by an i-event which goes before the e-event. But in Model 8.5 the resetting effect of an i-event is on the potential built up prior to its arrival. Since the characteristic property of the i-event here is to <u>reset</u> the potential, Model 8.5 and Model 8.6 below can be called <u>reset models</u>. This terminology will be carried over into Chapter 9 where similar models are considered.

<u>8.4.1</u> <u>Model 8.6</u> (Ten Hoopen and Reuver, 1967a; Srinivasan, Rajamannar and Rangan, 1971)

Consider a neuron with the following properties:

1) The e-events form a renewal process with pdf $f(\cdot)$, each e-event increasing the membrane potential $X(t)$ by one unit (the EPSP).

2) When $X(t)$ reaches N, the threshold, the neuron fires and is reset to the rest level, zero for convenience.

3) The i-events form a renewal process with pdf $g(\cdot)$, an i-event resetting $X(t)$ to zero.

4) The e-events are not independent of the i-events. An i-event triggers a sequence of e-events and stops the sequence started by the previous i-event.

The pdf of the interval between two sequences of firings can be found. However, it is obvious that the firing sequence is not a renewal process. Hence the interval pdf is not sufficient information, and higher order properties are needed to characterise the process. To this end the quantities $\pi(\cdot)$ and $p_k(\cdot)$ as defined by Lemmas 8.3.4 and 8.3.6 will be used here.

Lemma 8.4.1

$$\pi(t) = \gamma \int_t^\infty g(w) \; h_{eN}(w-t) \; dw \tag{8.4.1}$$

where $h_{eN}(\cdot)$ is the renewal density of the N-fold convolute of the e-events and γ is a normalising constant such that

$$\int_0^\infty \pi(t) \, dt = 1 . \tag{8.4.2}$$

Lemma 8.4.2 If

$$\Psi_k(t) = \lim_{\Delta,\Delta' \to 0} \text{Prob[the k-th i-event after the origin occurs in } (t,t+\Delta') \mid \text{a firing occurs in } (-\Delta,0)]/\Delta'$$

then

$$\Psi_1(t) = \pi(t) \int_t^\infty f_N(\tau) \, d\tau \tag{8.4.4}$$

and, for $k > 1$,

$$\Psi_k(t) = \int_0^t g(v) \left\{ \int_v^\infty f_N(\tau) \, d\tau \right\} \Psi_{k-1}(t-v) \, dv. \tag{8.4.5}$$

Lemma 8.4.3

$$p_0(t) = f_N(t) \int_t^\infty \pi(v) \, dv \tag{8.4.6}$$

and, for $k \geqslant 1$,

$$p_k(t) = \int_0^t f_N(v)\left\{\int_v^\infty g(\tau)\, d\tau\right\}\psi_k(t-v)\, dv. \tag{8.4.7}$$

Using these lemmas an expression for the interval pdf **is derived.**

Theorem 8.4.1. The interval pdf $P(\cdot)$ in Model 8.6 is given by

$$P(t) = \sum_{k=0}^{\infty} p_k(t). \tag{8.4.8}$$

Obviously the firings do not form a renewal process if the i-events do not form a Poisson process. Higher order properties are therefore to be known to provide more information than the pdf can. Thus the first order product density of the firings is given by

Theorem 8.4.2. The first order product density of the firings in Model 8.6 is given by

$$h_1(t) = h_{eN}(t)\int_t^\infty \pi(u)\, du + \int_0^t F(u)\chi(t-u)\, h_{eN}(t-u)\, du, \tag{8.4.9}$$

where $F(\cdot)$ and $\chi(\cdot)$ are given by

$$F(t) = \pi(t) + \int_0^t \pi(u)\, h_i(t-u)\, du, \tag{8.4.10}$$

$h_i(\cdot)$ being the renewal density of the i-events, and

$$\chi(t) = \int_t^\infty g(x)dx. \tag{8.4.11}$$

8.5 Discussion

In this chapter a large number of models in which the membrane potential occupies only discrete levels have been discussed. Thus many kinds of neurons can be modelled using these formulations. However, it is quite clear that in almost all of them an excessive amount of computation is needed. Further the very nature of the modelling approach precludes any meaningful inclusion of the decay parameter in the model. The 'decay' mechanism in Model 8.4 is at best an artifice, this is discussed in Section 10.1. The assumption of unitary changes in the membrane potential due to input impulses is not quite valid, because the change is a random variable with a size distribution (see p. 11). Thus the neuron does not fire for exactly the same number of impulses and there is wide variability in the firing pattern, though Penner (1972) discusses a model for an auditory neuron in which detection in the presence of noise is based on the summation of the number of pulses rather than of their energy. One way of taking into account the random size of the quantal changes is to use transition probabilities corresponding to the quantal size distribution in equation (8.1.29) and reformulate the neuron model accordingly.

Recently, Griffith (1971, p. 23) has observed that doubts have arisen about the quantal nature of the PSPs, though it is not clear what he meant by this remark. It is possible to consider the PSP to be a continuously distributed r.v. in neuron models. This has led to several continuous state models of neuron spike discharge which are discussed in the next chapter.

CHAPTER 9

CONTINUOUS STATE MODELS

It has been mentioned before that experiments have shown that the post-synaptic changes in potential (EPSP and IPSP) have a distribution in size. The membrane potential therefore does not change in steps of constant size but executes a random walk in continuous state space. This property introduces further variability into the spontaneous activity of neurons. The membrane potential therefore cannot be considered a discrete process like the birth-and-death process; the state space is now continuous. Such processes have been studied in recent years because of their applications in many areas. Cox (1962), for example, has studied a continuous state process, which he calls a cumulative process, that is associated with a renewal process, and solved the first passage time problem relating to this process. However, in a neuron there are two types of changes of state due to two input sequences - excitatory and inhibitory - and hence the associated cumulative process is more difficult to study. Nevertheless, useful continuous state models of neuronal firings have been recently formulated which can yield tangible results when simplifying assumptions are made. In this chapter such models are described in detail. Before introducing the models, the cumulative process of Cox (1962) is discussed in Section 9.1. This is the basis on which the models are built. Since these models are more or less continuous state analogues of discrete state models in the previous chapter, the development here is on parallel lines. Thus models with only one type of input are discussed in Section 9.2, models with independent streams of e-and i-events in Section 9.3 and models with dependent interaction of e- and i-events in Section 9.4.

9.1 Cumulative processes

Consider an ordinary renewal process with interval pdf $f(\cdot)$, and associate with the i-th renewal after the origin a continuous r.v. W_i. Let the W_is be iid with pdf $q(\cdot)$ and independent of the renewal sequence. Define a r.v. $X(t)$:

$$X(t) = \sum_{i=1}^{N(t)} W_i, \quad N(t) = 1,2,\ldots.$$

$$= 0 \quad , \quad N(t) = 0, \tag{9.1.1}$$

where $N(t)$ is the number of renewals in $(0,t)$ in the ordinary renewal process. The process $\{X(t)\}$ is called a <u>cumulative process</u>.

9.1.1 The distribution of X(t)

Define the distribution of $X(t)$ as

$$h(x,t) = \lim_{\Delta \to 0} \frac{[x < X(t) < x + \Delta]}{\Delta}. \tag{9.1.2}$$

Theorem 9.1.1 The distribution function $h(x,t)$ of the stochastic process $X(t)$ is given by

$$h(x,t) = \sum_{k=1}^{\infty} q_k(x) \int_0^t f_k(u)\, du \int_{t-u}^{\infty} f(v)\, dv$$

$$+ \delta(x) \int_t^{\infty} f(v)\, dv, \tag{9.1.3}$$

where the subscript stands for convolution of that order and $\delta(\cdot)$ is the delta function.

The proof is simple. $X(t)$ equals x if it reaches x after k inputs, $k \geqslant 1$. The probability density of this is the first term. Then there is an additional possibility of $X(t)$ remaining at zero without receiving any input. This accounts for the second term.

In Cox's theory, the r.v.s W_i are, in general, distributed over $(-\infty, \infty)$ so that an input can be negative or positive. Hence defining the two-sided Laplace transform as

$$^*q(p) = \int_{-\infty}^{\infty} q(x)\, e^{-px}\, dx, \tag{9.1.4}$$

the double Laplace transform of $h(\cdot,\cdot)$ is obtained as

$$^{*}h^{*}(p,s) = \frac{1 - f^{*}(s)}{s[1 - {}^{*}q(p)\, f^{*}(s)]} \, . \qquad (9.1.5)$$

The structure of this equation precludes an explicit inverse for $^{*}h^{*}(p,s)$ except when $q(\cdot)$ and $f(\cdot)$ have a simple form. An example may be considered. Let the renewal sequence be a Poisson process with parameter λ ,
i.e.,

$$f(t) = \lambda e^{-\lambda t} \, . \qquad (9.1.6)$$

Then,

$$^{*}h(p,t) = \exp\,[-\lambda t\,(1 - {}^{*}q(p))]. \qquad (9.1.7)$$

If the input W is exponentially distributed, i.e.,

$$\begin{aligned} q(x) &= \alpha e^{-\alpha x}, \quad x \geqslant 0 \\ &= 0 \quad , \quad x < 0, \end{aligned} \qquad (9.1.8)$$

then

$$^{*}q(p) = \frac{\alpha}{\alpha + p} \qquad (9.1.9)$$

and $^{*}h(p.t)$ is written as

$$^{*}h(p,t) = e^{-\lambda t} + e^{-\lambda t}\left\{ \exp\left(\frac{\lambda t\alpha}{\alpha + p}\right) - 1 \right\} . \qquad (9.1.10)$$

The inverse of this is

$$h(x,t) = e^{-\lambda t}\,\delta(x) + \sqrt{\frac{\lambda t\alpha}{x}}\; e^{-\lambda t - \alpha x}\, I_1\left\{ 2\sqrt{\lambda t\alpha x} \right\} , \qquad (9.1.11)$$

where the first term is a concentration at x = 0, and in the second, $I_1(\cdot)$ is the modified Bessel function of the first order.

The above discussion is for an unrestricted random walk with x ranging over all values in $(-\infty, \infty)$. However, since the neuron membrane potential is restricted above by the threshold, the more relevant problem is the first passage time problem.

9.1.2 The first passage time problem in a cumulative process.

For simplicity, W is assumed to be a positive valued r.v. Then X(t) is non-decreasing. Let the r.v. representing the time when X(t) crosses a level K for the first time be T.
Then

$$T > t \quad \text{iff} \quad X(t) < K. \tag{9.1.12}$$

The pdf $P(\cdot)$ of the first passage time T is now derived.

Lemma 9.1.1

$$\int_t^\infty P(u)\,du = \int_0^K h(x,t)\,dx. \tag{9.1.13}$$

This follows from (9.1.12).

On taking the Laplace transform of (9.1.13) w.r.t. t,

$$P^*(s) = 1 - s \int_0^K h^*(x,s)\,dx. \tag{9.1.14}$$

To obtain more explicit results, a Laplace transform w.r.t. K is taken

$$\tilde{P}^*(s) = \frac{1}{p_1} - \frac{s}{p_1}\,\tilde{h}^*(p_1,s) \tag{9.1.15}$$

which on using (9.1.5) with a slight difference in notation (due to K being defined in $(0,\infty)$ rather than $(-\infty, \infty)$) leads to

Theorem 9.1.2. The double Laplace transform w.r.t. t and the threshold K of the first passage time in the cumulative process is given by

$$\tilde{P}^*(s) = \frac{f^*(s)\,[1 - \tilde{q}(p_1)]}{p_1[1 - \tilde{q}(p_1)\, f^*(s)]} . \tag{9.1.16}$$

$P(\cdot)$ is obtained by inverting (9.1.16) w.r.t. K as well as t. Exact results can be obtained assuming $q(\cdot)$ and $f(\cdot)$ to be Erlangian distributions. For example, let $f(\cdot)$ be given by (9.1.6) and $q(\cdot)$ by (9.1.8). Then

$$\tilde{P}^*(s) = \frac{\lambda}{p\lambda + s\alpha + ps} . \tag{9.1.17}$$

Inverting w.r.t. s,

$$\tilde{P}(t) = \frac{\lambda}{p + \alpha} \exp\left(- \frac{p\lambda t}{p+\alpha}\right) \tag{9.1.18}$$

$$= e^{-\lambda t} \sum_{m=0}^{\infty} \frac{(\alpha t)^m}{m!} \left(\frac{\lambda}{p + \alpha}\right)^{m+1} , \tag{9.1.19}$$

which on inverting w.r.t. K, gives

$$P(t) = \lambda e^{-\lambda t} e^{-\alpha K} I_0\left\{ 2\sqrt{\lambda t \alpha K} \right\} , \tag{9.1.20}$$

$I_0(\cdot)$ being the modified Bessel function of order zero.

The first passage time problem described above can be approached in a slightly different but none the less instructive way. First $h(\cdot,\cdot)$ is redefined as below.

$$h(x,t) = \lim_{\substack{\Delta \to 0 \\ \Delta_1 \to 0}} \text{Prob}\,[X(t) < x < X(t+\Delta) < x+\Delta_1] / \Delta\Delta_1 . \tag{9.1.21}$$

Notice that $h(x,t)$ now gives the distribution of $X(t)$ immediately after an input due to a primary event (of the underlying renewal process) occurs. Then it is easy to arrive at

<u>Lemma 9.1.2</u> The distribution function $h(\cdot,\cdot)$ as defined in (9.1.21) is given by

$$h(x,t) = f(t)\, q(x) + \int_0^t f(t-u)\, du \int_0^x h(x',u)\, q(x-x')\, dx'. \tag{9.1.22}$$

The first passage time pdf is then given by

<u>Theorem 9.1.3</u>. The first passage time in a cumulative process with upper boundary K, the interval between successive events in the underlying renewal process having pdf $f(\cdot)$ and the increments due to these events being iid positive valued r.v.s with pdf $q(\cdot)$, has the pdf

$$P(t) = f(t) \int_K^\infty q(x)\, dx + \int_0^t f(t-u)\, du \int_0^K h(x,u)\, dx \int_{K-x}^\infty q(x')\, dx'. \tag{9.1.23}$$

Yet another form of the pdf of the first passage time is given by

<u>Theorem 9.1.4</u>.

$$P(t) = f(t) \int_K^\infty q(x)\, dx + \sum_{n=2}^\infty f_n(t) \int_0^K q_{n-1}(x)\, dx \int_{K-x}^\infty q(x')\, dx'. \tag{9.1.24}$$

On taking the double Laplace transform of (9.1.23) and (9.1.24) w.r.t. t and K, the result is (9.1.16). While (9.1.24) is simple and straightforward, the approach using Lemma 9.1.2 and Theorem 9.1.3 seems roundabout, but this is to illustrate a method which will be used in the analysis in the sections to follow.

The cumulative process is obviously similar to the process of build-up of the membrane potential in a neuron. Now in the case of the neuron, when $X(t)$ crosses the threshold level K, it is immediately reset to the rest level (assumed zero) so that this cumulative process repeats itself, 'breaking down' whenever it crosses K, to start again from the rest state. This process is called the c-process. Thus the c-process is the sequence of K-crossings or firings generated by the cumulative process of Cox. Since the primary events form a renewal process, the c-process is also a renewal process. In the following sections several models based on the c-process are discussed. The first continuous state model was proposed by Osaki and Vasudevan (1972). This is merely Model 8.5 in Chapter 8 with the only difference that the excitatory post-synaptic changes are not constant but are r.v.s with common pdf $q(\cdot)$. As observed in Section 8.3.3, the sequence of firings is not a renewal process. Osaki and Vasudevan (1972) did not take care of this fact and were consequently led to some results that did not correspond to the model proposed by them. This model has been studied in detail by Srinivasan (1976), who has derived higher order properties of the sequence, and Srinivasan and Sampath (1976) who have obtained an expression for the pdf of the first passage time of the sequence of firings. In another model, Srinivasan and Sampath (1975) have assumed the sequence of undeleted excitatories resulting from pre-synaptic interaction (see Chapter 5) to give rise to EPSPs that are iid r.v.s with pdf $q(\cdot)$ along with linear decay of the membrane potential to the rest level in the absence of inputs. Other models in which the sequence of inhibitories is controlled by the excitatory impulses are also discussed. Finally a model in which the input sequences are Poisson streams and the membrane potential decays linearly to the rest level in the absence of inputs both above and below the threshold level is presented.

9.2 Models with only one input sequence (cf. Section 8.2)

9.2.1 Model 9.1

This is the basic model of the c-process discussed in the last section. Consider the neuron to have the following properties:

1) The neuron receives a sequence of e-events which form a renewal process with interval pdf $f(\cdot)$.
2) Each e-event increases the post-synaptic membrane potential $X(t)$ by an amount — the EPSP — that is a r.v. Successive e-events cause increases that are iid with pdf $q(\cdot)$.
3) When $X(t)$ crosses a threshold level K the neuron fires immediately after which $X(t)$ is set to the rest level (zero, for convenience).

Theorem 9.2.1 The firings in Model 9.1 form a renewal process with the interval between two successive firings having a pdf $P(\cdot)$ given by Theorem 9.1.4.

A specific example may be considered now. Let the e-events form a Poisson process with parameter λ, i.e.,

$$f(t) = \lambda e^{-\lambda t},$$

and let the post-synaptic increment distribution $q(\cdot)$ be given by

$$\begin{aligned} q(x) &= \alpha^2 x e^{-\alpha x}, \quad x \geqslant 0 \\ &= 0 \quad\quad\quad\;\;, \quad x < 0. \end{aligned} \tag{9.2.1}$$

This can be used to approximate a distribution of the kind obtained by Katz and his colleagues for the miniature end plate potentials in the neuromuscular junction (Knight, 1972a, p. 736). Then the Laplace transform of the pdf of the firing sequence is obtained from (9.1.24):

$$\begin{aligned} P^*(s) = \sqrt{f^*(s)}\, e^{-\alpha K} [&\sinh (\alpha K \sqrt{f^*(s)}) \\ &+ \sqrt{f^*(s)}\; \cosh (\alpha K \sqrt{f^*(s)})]. \end{aligned} \tag{9.2.2}$$

9.2.2 Model 9.2 (Srinivasan and Sampath, 1975)

The decay of the membrane potential in the absence of inputs is an important property of the membrane. On page 10 it has been stated that this natural decay is exponential. The addition of such a variation to the c-process leads to non-linearities in the defining equations that do not allow of analytic solution. However, it is possible to use a linear approximation of this natural decay in the c-process; this leads to tangible results. In fact the only kind of decay that leads to an analytic solution in a continuous state process is the linear. The present model which uses only one kind of input sequence, may be considered a model of a neuron which has pre-synaptic interaction of e-and i-events (Chapter 5) so that the post-inhibition sequence is a single sequence of r-events which cause EPSPs that are integrated by the membrane. Hence this model, though it is built on only one sequence of inputs, may be considered an extension of the deletion models of Chapter 5. This, incidentally, effectively removes a defect of deletion models (they do not consider the evolution of membrane potential with time). The model has the following properties:

1) The e-events form a renewal process with pdf $\varphi(\cdot)$.
2) The i-events form a Poisson process with parameter μ. (This ensures that the sequence of undeleted e-events is a renewal process).
3) An i-event deletes the first e-event following it.
4) An undeleted e-event (r-event) increases the membrane potential $X(t)$ by a random amount. Successive r-events cause increases (EPSPs) that are iid with pdf $q(\cdot)$.
5) In the absence of an r-event, $X(t)$ decays linearly with time to zero level at a rate $\underline{b}$ units per second.
6) When $X(t)$ crosses the threshold level K, the neuron fires and $X(t)$ is reset to zero immediately.

Since the i-events are Poisson, the r-events form a renewal process with pdf $f(\cdot)$ whose Laplace transform $f^*(s)$ is given by Theorem 5.1.2. If the i-events are not Poisson, the r-events do not form a renewal process (Section 5.1.2) and the analysis becomes extremely complicated. Due to the decay, the membrane potential is somewhat like the water level in a finite dam (Srinivasan, 1974b). Since the inputs due to the r-events are positive $X(t)$ is always ≥ 0 and is bounded above by K. Hence $X(t)$ is a simple stochastic process on $(0,K)$. Since a firing coincides with an r-event, the firing sequence is a renewal process. Hence the first passage time distribution, i.e., the pdf of the interval between two successive firings, completely describes the firing sequence. To obtain this, $h(\cdot,\cdot)$ is appropriately redefined in the context of the firing process.

Let

$$h(x,t) = \lim_{\Delta,\Delta',\Delta''\to 0} \text{Prob}\,[X(t) < x < X(t+\Delta) < x+\Delta' \mid 0 < X(u) < K,\ \forall\, u \in (0,t),\ \text{a firing in } (-\Delta'',0)]/\Delta\,\Delta'. \tag{9.2.3}$$

<u>Lemma 9.2.1</u>

$$h(x,t) = f(t)\,q(x) + \int_{\max(0,t-K/b)}^{t} f(t-u)\,du \int_{b(t-u)}^{\min(K,x+b(t-u))} h(x',u)\,q[x-(x'-b(t-u))]dx' + q(x)\int_0^t f(t-u)\,du \int_0^{\min(K,b(t-u))} h(x',u)\,dx'. \tag{9.2.4}$$

The lemma follows from the fact that an input can be due to the first r-event or due to a subsequent r-event which can occur before or after $X(t)$ has decayed to zero. The minimum in the upper limit restricts $X(t)$ to K and the maximum restricts the time t to be positive, as it should be.

The pdf of the first passage time, $P(\cdot)$, is given by

<u>Theorem 9.2.2</u> The firings in Model 9.2 form a renewal process with pdf $P(\cdot)$ given by

$$P(t) = f(t)\int_K^{\infty} q(x)\,dx + \int_{\max(0,t-K/b)}^{t} f(t-u)\,du$$

$$\int_{b(t-u)}^{K} h(x,u)\, dx \int_{K-(x-b(t-u))}^{\infty} q(y)\, dy$$

$$+ \int_{K}^{\infty} q(x)\, dx \int_{0}^{t} f(t-u)\, du \int_{0}^{\min(K,b(t-u))} h(x,u)\, dx. \qquad (9.2.5)$$

The theorem is proved by observing that a firing can occur due to (a) the first r-event or (b) a subsequent r-event before $X(t)$ has decayed to zero or (c) the first r-event after $X(t)$ has decayed to zero. Equation (9.2.5) is written taking into account these three mutually exclusive and exhaustive sets of events.

To solve the integral equations in (9.2.4) and (9.2.5), first define the double Laplace transform of $h(x,t)$ as

$$\bar{h}^*(p,s) = \int_{0}^{K} e^{-px}\, dx \int_{0}^{\infty} e^{-st}\, h(x,t)\, dt, \qquad (9.2.6)$$

and

$$\bar{q}(p) = \int_{0}^{K} e^{-py}\, q(y)\, dy. \qquad (9.2.7)$$

After considerable simplification one can arrive at

<u>Lemma 9.2.2</u>

$$\bar{h}^*(p,s) = \bar{q}(p)\, f^*(s)$$

$$+ \int_{0}^{K} h^*(x,s)\, e^{-px}\, dx \int_{0}^{x/b} f(u)\, e^{-(s-pb)u}\, du$$

$$\cdot \int_{0}^{K+bu-x} q(y)\, e^{-py}\, dy$$

$$+ \bar{q}(p) \int_0^K h^*(x,s)\, dx \int_{x/b}^{\infty} f(u)\, e^{-su}\, du. \qquad (9.2.8)$$

Transforming (9.2.5) leads to

<u>Theorem 9.2.3</u> The interval between two successive firings in Model 9.2 has a pdf $P(\cdot)$ whose Laplace transform is given by

$$P^*(s) = f^*(s)\,[1 - \bar{q}\,(p)]$$

$$+ \int_0^K h^*(x,b)\, dx \int_0^{x/b} f(u)\, e^{-su}\, du \int_{K+bu-x}^{\infty} q(y)\, dy$$

$$+ [1-\bar{q}\,(p)] \int_0^K h^*(x,s)\, dx \int_{x/b}^{\infty} f(u)\, e^{-su}\, du. \qquad (9.2.9)$$

To obtain an explicit formula for $P^*(s)$, one must solve for $\bar{h}^*(p,s)$. At this stage it is necessary to assume that $f(\cdot)$ and $q(\cdot)$ have a simple form. Now there exists a wide class of distributions that can be approximated to any arbitrary degree of accuracy by a general Erlangian (Cox, 1962, p.16). For simplicity assume that

$$f(t) = \lambda(\lambda t)^{m-1}\, e^{-\lambda t}/(m-1)\,! \qquad (9.2.10)$$

and

$$q(x) = \alpha\,(\alpha x)^{n-1}\, e^{-\alpha x}/(n-1)!\,. \qquad (9.2.11)$$

Then

$$\bar{q}(p) = [\alpha/(\alpha+p)]^n \left(1 - \sum_{j=0}^{n-1} ((\alpha+p)\,K)^j\, e^{-(\alpha+p)K}/j\,! \right), \qquad (9.2.12)$$

and

$$f^*(s) = (\lambda/(\lambda+s))^m. \qquad (9.2.13)$$

Since a general Erlangian is a weighted sum of the special Erlangian in (9.2.10) and (9.2.11), the extension is obvious. Then (9.2.8) becomes

$$\bar{h}^*(p,s) = \bar{q}(p)\Big(\lambda/(\lambda+s)\Big)^m\Big(1 + \sum_{j=0}^{m-1} (\lambda+s)^j T_j\Big)$$

$$+ \Big(\alpha/(\alpha+p)\Big)^n \Big(\lambda/(\lambda+s-pb)\Big)^m \cdot$$

$$\Big\{ \bar{h}^*(p,s) - \sum_{j=0}^{m-1} (\lambda+s-pb)^j T_j \Big\}$$

$$- \Big(\alpha/(\alpha+p)\Big)^n e^{-pK} \sum_{j=0}^{n-1} (\alpha+p)^j R_j \qquad (9.2.14)$$

where

$$T_j = \frac{1}{b^j j!} \int_0^K h^*(x,s)\, x^j e^{-(\lambda+s)x/b}\, dx \qquad (9.2.15)$$

and

$$R_j = \frac{e^{-\alpha K}}{j!} \int_0^K h^*(x,s)\, e^{\alpha x}\, dx$$

$$\int_0^{x/b} f(u)\, e^{-(s+b\alpha)u} (K+bu-x)^j\, du, \qquad (9.2.16)$$

and (9.2.9) gives

$$P^*(s) = f^*(s)\,[1-\bar{q}(p)]\Big\{1 + \sum_{j=0}^{m-1} (\lambda+s)^j T_j\Big\}$$

$$+ \sum_{j=0}^{n-1} \alpha^j R_j. \qquad (9.2.17)$$

Transposing (9.2.14),

$$\bar{h}^*(p,s) = (\alpha/(\alpha+p))^n \Big[\Big(1 - \sum_{j=0}^{n-1} ((\alpha+p)\,K)^j\, e^{-(\alpha+p)K}\Big)$$

$$\cdot \Big\{1 + \sum_{j=0}^{m-1} (\lambda+s)^j\, T_j\Big\}\Big(\lambda/(\lambda+s)\Big)^m$$

$$- \Big(\lambda/(\lambda+s-pb)\Big)^m \sum_{j=0}^{m-1} (\lambda+s-pb)^j\, T_j$$

$$- e^{-pK} \sum_{j=0}^{n-1} (\alpha+p)^j\, R_j \Big] / [1 - \big(\alpha/(\alpha+p)\big)^n$$

$$\Big(\lambda/(\lambda+s-pb)\Big)^m]. \qquad (9.2.18)$$

It is now observed that the T_js and R_js are all independent of p and hence constants as far as p is concerned. Since $\bar{h}^*(p,s)$ is an entire function of p, the numerator in (9.2.18) must vanish at all the values of p for which the denominator is zero. The denominator has m+n zeros. At these m+n values of p, the condition on the numerator of (9.2.18) gives m+n simultaneous equations relating the m+n constants, viz., T_j, $j = 0,1,\cdots,m-1$ and R_j, $j = 0,1,\cdots, n-1$. Thus $P^*(s)$ in (9.2.17) is fully determined.

However, the denominator is a polynomial in p, the zeros of which can be determined analytically only upto the fourth degree. Therefore if an analytic solution of $P^*(s)$ is desired, the sum of the highest orders of the Erlangian distributions used for $f(\cdot)$ and $q(\cdot)$ should not exceed 4. To illustrate, let $f(\cdot)$ and $q(\cdot)$ be exponential, i.e., they are given by (9.2.10) and (9.2.11) with m=1 and n=1 respectively. Then (9.2.18) becomes

$$\bar{h}^*(p,s) = [(\alpha/(\alpha+p))\,(1-e^{-(\alpha+p)K})(1+T_0)\,(\frac{\lambda}{\lambda+s})$$

$$- (\lambda/(\lambda+s-pb))\,T_0 - e^{-pK}\, R_0]/$$

$$[1 - (\alpha/(\alpha+p))\,(\lambda/(\lambda+s-pb))]. \qquad (9.2.19)$$

The roots of the denominator are

$$p_{1,2} = [\lambda + s - b\alpha \pm \sqrt{(\lambda + s - b\alpha)^2 + 4b\alpha s}]/2b. \qquad (9.2.20)$$

Then, from the denominator of (9.2.19),

$$T_0 = \frac{(\lambda+s)\,[(\lambda+s-p_2b)\,e^{-p_2K} + (\lambda+s-p_1b)\,e^{-p_1K}]}{[b(\lambda+s-p_2b)\,p_1e^{-p_1K} - b\,(\lambda+s-p_1b)\,p_2e^{-p_2K} - 1]} \qquad (9.2.21)$$

and

$$R_0 = \frac{[(p_1-p_2)\,b + (\lambda+s-p_1b)\,e^{-(\alpha+p_1)K} - (\lambda+s-p_2b)\,e^{-(\alpha+p_2)K}]\,\lambda}{b(\lambda+s-p_2b)p_1\,e^{-p_1K} - b(\lambda+s-p_1b)\,p_2\,e^{-p_2K}} \qquad (9.2.22)$$

This leads to

<u>Theorem 9.2.4</u> In Model 9.2, if the e-events form a Poisson process with parameter λ and the post synaptic increments are iid r.v.s exponentially distributed with parameter α, then the Laplace transform of the pdf of the interval between two successive firings is

$$P^*(s) = (\lambda/(\lambda+s))\,e^{-\alpha K}\,(1 + T_0) + R_0. \qquad (9.2.23)$$

<u>Corollary</u> The mean and the variance of the firing interval T are given by

$$\bar{T} = [\lambda(\lambda-b\alpha)(1+\alpha K) - \lambda b\alpha + b^2\alpha^2 \cdot \exp(-(\lambda-b\alpha)K/b)]/(\lambda(\lambda-b\alpha)^2),\ \lambda \neq b\alpha$$

$$= (1 + \alpha K + \alpha^2K^2/2)/\lambda,\ \lambda = b\alpha \qquad (9.2.24)$$

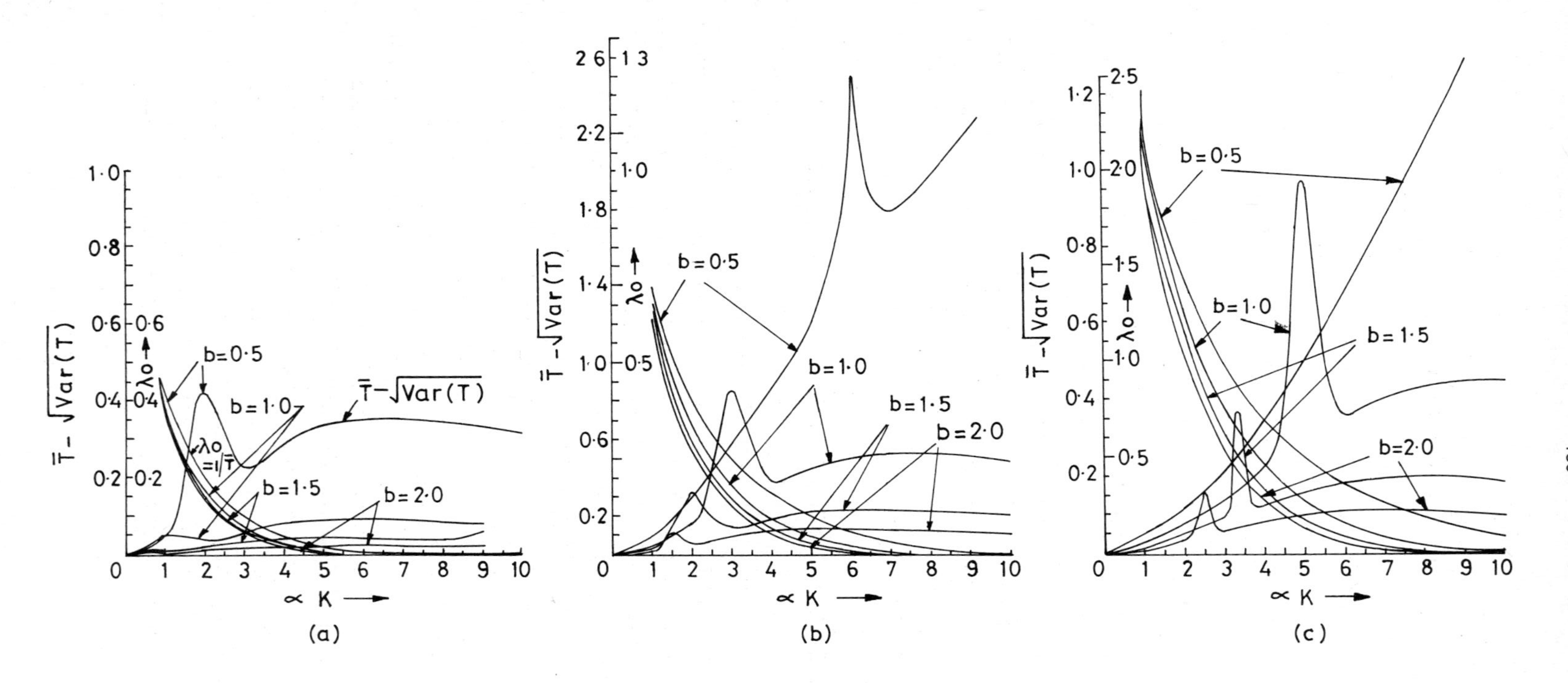

FIG. 9·2·1 GRAPHS OF $1/\overline{T}$ AND $\overline{T} - \sqrt{Var(\tau)}$ vs K

(a) $\lambda = 1$

(b) $\lambda = 3$

(c) $\lambda = 5$

and

$$\mathrm{Var}\,(T) = [\,\lambda^2(\lambda-b\alpha)^2 - 2\,\lambda^3 b\alpha - 3\,\lambda^2 b^2\alpha^2$$

$$+ 2\,\lambda^3(\lambda-b\alpha)\alpha K + (8\,\lambda^2 b^2\alpha^2 - 4\lambda b^3\alpha^3$$

$$+ 2\lambda b\alpha\,(\lambda^2-b^2\alpha^2)\,\alpha K)\,\exp\,(-(\lambda-b\alpha)\,K/b)$$

$$+ b^4\alpha^4\,\exp\,(-(\lambda-b\alpha)\,2K/b)\,]/\,\lambda^2(\lambda-b\alpha)^4,\ \lambda \neq b\alpha$$

$$= (1+2\alpha K + 2\alpha^2K^2 + \alpha^3K^3)/\lambda^2,\quad \lambda = b\alpha. \qquad (9.2.25)$$

Graphs relating $1/\bar{T}$ and $\bar{T} - \sqrt{\mathrm{Var}(T)}$ to αK for different values of b and λ are given in Figure 9.2.1. Here the threshold K is normalised to unity so that αK is a measure of the mean number of inputs required for firing. The significance of the $\bar{T} - \sqrt{\mathrm{Var}\,(T)}$ graph is discussed in Section 10.3.

9.3 Models with independent interaction of e-and i-events

A neuron has the outputs of several other neurons converging on it; even one incoming axon sometimes branches out and makes several synaptic connections on the dendrites and the soma of the neuron. Some of them have an excitatory effect, others inhibitory. The effect may be pre-synaptic or post-synaptic. In the last section only pre-synaptic interaction of e- and i-events leading to a single sequence of r-events has been considered, the r-events increasing the post-synaptic membrane potential by random amounts. Since the inhibitory effect can also be post - synaptic (leading to an IPSP), a more realistic model must consider both excitatory and inhibitory post-synaptic effects. If the number of synapses is large, then the e-events and the i-events can both be considered Poisson processes (resulting from superposition). A model with Poisson input sequences is considered below. In this the hyper-polarising effect (i.e., the membrane potential going below the resting level) due to IPSPs is taken into account along with the decay of the membrane potential to the rest level when it is both above and below the rest level. Another kind of model is the continuous state analogue of Model 8.5. This extension is due to

Osaki and Vasudevan (1972). This model is studied in detail in Section 9.3.2 following Srinivasan (1976) and Srinivasan and Sampath (1976).

9.3.1 Model 9.3

This model has the following properties:

1) The e-events and the i-events both form Poisson processes with parameter λ and μ respectively.
2) An e-event increases the membrane potential $X(t)$ by an EPSP that is a r.v., successive such increases being iid with pdf $q(\cdot)$.
3) An i-event, similarly, decreases $X(t)$ by an IPSP which has a pdf $r(\cdot)$.
4) In the absence of inputs, $X(t)$ decays to the rest level (zero) at a rate b units / sec when $X(t) > 0$ and c units / sec when $X(t) < 0$.
5) When $X(t)$ crosses the threshold level K, the neuron fires and $X(t)$ is reset to the rest level.

Define the following distributions:

$$h(x,t) = \lim_{\Delta,\Delta',\Delta''\to 0} \text{Prob}\,[(X(t) < x < X(t+\Delta') < x + \Delta'') \cup (X(t) > x > X(t+\Delta') > x - \Delta'') \mid 0 < X(u) < K,\ \forall u \in (0,t), \text{ a firing in } (-\Delta,0)]/\Delta'\Delta'' \tag{9.3.1}$$

and

$$g(z,t) = \lim_{\Delta,\Delta',\Delta''\to 0} \text{Prob}\,[(|X(t)| < z < |X(t+\Delta')| < z + \Delta'') \cup (|X(t)| > z > |X(t+\Delta')| > z - \Delta'') \mid 0 < X(u) < \infty,\ \forall u \in (0,t), \text{ a firing in } (-\Delta,0)]/\Delta'\Delta'', \tag{9.3.2}$$

The two distributions $h(x,t)$ and $g(z,t)$ are thus defined for membrane potentials above and below the rest level respectively.

Lemma 9.3.1

$$h(x,t) = q(x)\lambda e^{-(\lambda+\mu)t} +$$

$$\int_{\max(0,t-K/b)}^{t} \lambda e^{-(\lambda+\mu)(t-u)} du \int_{b(t-u)}^{\min(K,x+b(t-u))} h(y,u)\, q(x-(y-b(t-u)))\, dy$$

$$+ q(x) \int_0^t \lambda e^{-(\lambda+\mu)(t-u)} du \int_0^{\min(K,b(t-u))} h(y,u)\, dy$$

$$+\int_{\max(0,t-(K-x)/b)}^{t} \mu e^{-(\mu+\lambda)(t-u)} du \int_{x+b(t-u)}^{K} h(y,u)\, r(y-b(t-u)-x)\, dy$$

$$+ q(x) \int_0^t \lambda e^{-(\lambda+\mu)(t-u)} du \int_0^{c(t-u)} g(z,u)\, dz$$

$$+ \int_0^t \lambda e^{-(\lambda+\mu)(t-u)} du \int_{c(t-u)}^{\infty} g(z,u)\, q(x+z-c(t-u))\, dz. \qquad (9.3.3)$$

Lemma 9.3.2

$$g(z,t) = r(z)\mu e^{-(\mu+\lambda)t} +$$

$$\int_0^t \mu e^{-(\mu+\lambda)(t-u)} du \int_{c(t-u)}^{z+c(t-u)} g(y,u)\, r(z-(y-c(t-u)))\, dy$$

$$+ r(z) \int_0^t \mu e^{-(\mu+\lambda)(t-u)} du \int_0^{c(t-u)} g(y,u)\, dy$$

$$+ \int_0^t \lambda e^{-(\lambda+\mu)(t-u)} du \int_{z+c(t-u)}^{\infty} g(y,u)\, q(y-c(t-u)-z)\, dy$$

$$+ \; r(z) \int_0^t \mu e^{-(\mu+\lambda)(t-u)} du \int_0^{\min(K,b(t-u))} h(y,u)\, dy$$

$$+ \int_{\max(0,t-K/b)}^{t} \mu e^{-(\mu+\lambda)(t-u)} du \int_{b(t-u)}^{K} h(y,u)\, r(z+y-b(t-u))\, dy. \tag{9.3.4}$$

The pdf $P(\cdot)$ of the interval between two successive firings is then given by

<u>Theorem 9.3.1</u> The pdf $P(\cdot)$ of the interval between two successive firings in Model 9.3 is given by

$$P(t) = \lambda e^{-(\lambda+\mu)t} \int_K^{\infty} q(x)\, dx$$

$$+ \int_{\max(0,t-K/b)}^{t} \lambda \; e^{-(\lambda+\mu)(t-u)} du \int_{b(t-u)}^{K} h(x,u)\, dx \int_{K-(x-b(t-u))}^{\infty} q(x)\, dx$$

$$+ \int_K^{\infty} q(x)\, dx \int_0^t \lambda e^{-(\lambda+\mu)(t-u)} du \left\{ \int_0^{\min(K,b(t-u))} h(x,u)\, dx \right.$$

$$\left. + \int_0^{c(t-u)} g(z,u)\, dz \right\}$$

$$+ \int_0^t \lambda e^{-(\lambda+\mu)(t-u)} \, du \int_{c(t-u)}^{\infty} g(z,u)\, dz$$

$$\int_{K+z-c(t-u)}^{\infty} q(x)\, dx. \tag{9.3.5}$$

The proof of the above results is similar to the proof of (9.2.4) and (9.2.5) and is omitted. The procedure of solving for $P^*(s)$ is also similar to that in Section 9.2.2 with slight differences. Thus define the double Laplace transform of h(x,t) as

$$\bar{h}^*(p,s) = \int_0^K e^{-px}\, dx \int_0^\infty e^{-st}\, h(x,t)\, dt. \tag{9.3.6}$$

and

$$\tilde{g}^*(p_1,s) = \int_0^\infty e^{-p_1 z}\, dz \int_0^\infty e^{-st}\, g(z,t)\, dt. \tag{9.3.7}$$

As in Section 9.2.2, q(·) and r(·) are assumed to be approximated by a suitable combination of Erlangian distributions. For simplicity assume

$$q(x) = \alpha\, e^{-\alpha x}, \tag{9.3.8}$$

$$r(x) = \beta\, e^{-\beta x}. \tag{9.3.9}$$

Then

$$\bar{q}(p) = \frac{\alpha}{\alpha+p}\, [1 - e^{-(\alpha+p)K}] \tag{9.3.10}$$

and

$$\tilde{r}\,(p_1) = \frac{\beta}{\beta + p_1}\,. \tag{9.3.11}$$

Using these, after much simplification,

$$\bar{h}^*(p,s) = [\bar{q}(p)\ (1 + H_{00} + G_{00})\, \lambda\ /(\lambda + \mu + s)$$

$$- \frac{\alpha}{\alpha+p} \cdot H_{00}\, \lambda/(\lambda + \mu + s - pb) - e^{-pK}\, F_0$$

$$+ \frac{\beta\mu}{pb-(\lambda+\mu+s)} \Big\{ H_{00}/(\beta-(\lambda+\mu+s)/b)$$

$$+ \quad H_0 \left(\frac{1}{\beta-p} - \frac{1}{\beta - \frac{\lambda+\mu+s}{b}}\right) \Big\}$$

$$+ \quad \frac{\alpha}{\alpha+p} \left(1 - e^{-(\alpha+p)K}\right) \quad \frac{\lambda}{c\alpha - (\lambda+\mu+s)}$$

$$\left\{ G_{00} - G_0 \right\} \;]/[1 - \frac{\lambda}{\lambda+\mu+s-pb} \left(\frac{\alpha}{\alpha + p}\right)$$

$$+ \quad \frac{\mu}{pb - (\lambda+\mu+s)} \quad \left(\frac{\beta}{\beta - p}\right)] \,, \qquad (9.3.12)$$

and

$$\tilde{g}^*(p_1,s) = [\frac{\beta}{\beta + p_1}(1 + G_{00} + H_{00})$$

$$(\lambda+\mu+s+bp_1) \; \mu/(\lambda+\mu+s)$$

$$+ \left(\frac{\beta}{\beta + \frac{\lambda+\mu+s}{c}}\right) G_{00} \; (\lambda+\mu+s+bp_1) \; \mu/(cp_1-(\lambda+\mu+s))$$

$$+ \frac{(\lambda+\mu+s+bp_1}{cp_1-(\lambda+\mu+s)} \left\{ \left(\frac{\alpha}{\alpha - \frac{\lambda+\mu+s}{c}}\right) \quad G_{00}\right.$$

$$\left. + \; \alpha \; G_0 \left(\frac{1}{\alpha - p_1} \quad - \quad \frac{1}{\alpha - \frac{\lambda+\mu+s}{c}}\right)\right\}$$

$$+ \frac{\beta}{\beta + p_1}\mu \; H_0] \; / \; [(\lambda + \mu + s + bp_1)$$

$$\left\{ 1 + \frac{\beta}{\beta + p_1} \quad \frac{\mu}{cp_1 - (\lambda+\mu+s)} + \frac{\alpha}{\alpha - p_1} \; \frac{\lambda}{cp_1 - (\lambda+\mu+s)} \right\}]$$

$$(9.3.13)$$

where

$$H_{00} = \bar{h}^*(\frac{\lambda + \mu + s}{b}, s), \qquad (9.3.14)$$

$$G_{00} = \tilde{g}^*(\frac{\lambda + \mu + s}{c}, s), \qquad (9.3.15)$$

$$H_0 = \bar{h}^*(\beta, s), \qquad (9.3.16)$$

$$G_0 = \tilde{g}^*(\alpha, s), \qquad (9.3.17)$$

and

$$F_0 = \frac{e^{-\alpha K} \lambda}{\lambda + \mu + s + b\alpha} [\bar{h}^*(\alpha, s) - H_{00}]. \qquad (9.3.18)$$

<u>Theorem 9.3.3</u> The Laplace transform of the pdf of the interval between two successive firings when the post-synaptic increments and decrements are exponentially distributed with parameters α and β respectively is

$$P^*(s) = [1 - \frac{\alpha}{\alpha+p}(1 - e^{-\alpha K})](1 + H_{00} + G_{00}) \frac{\lambda}{\lambda+\mu+s} + F_0$$

$$+ e^{-\alpha K} \frac{\lambda}{c\alpha - (\lambda+\mu+s)} (G_{00} - G_0). \qquad (9.3.19)$$

This result involves a number of unknowns : G_0 , H_{00}, G_{00} and F_0. It is now observed that the quantities H_{00}, G_{00}, H_0, G_0 and F_0 are all independent of p and hence are constants. Next, $\bar{h}^*(p,s)$ is an entire function of p. Therefore at the values of p for which the denominator of (9.3.12) is zero the numerator of (9.3.12) must vanish. The denominator is a polynomial in p with 3 zeros, but the numerator has 5 unknowns (H_{00}, G_{00}, H_0, G_0 and F_0). To get a determinate solution, first observe that $\tilde{g}^*(p_1,s)$ is a meromorphic function of p_1. Since Re $p_1 > 0$, the function is analytic in the left half of the p_1 plane except at a finite number of **poles**. If there are any right half plane zeros in the denominator of (9.3.13), the numerator must vanish at those points. Now, the denominator of (9.3.13) is a polynomial of degree 4, while the numerator has 4 unknowns. It can be shown, using Rouché's theorem (see Copson, 1935), that there are exactly 2 zeros in the right half of the p_1 plane. The numerator of (9.3.13) must therefore vanish at these 2 zeros in the right half plane. This gives two more equations relating the five unknowns which together with the 3 equations obtained from (9.3.13) constitute 5 equations in 5 unknowns. The quantities H_{00}, G_{00}, G_0, H_0 and F_0 are thus known. Thus $P^*(s)$ is determined.

Corollary If in the analysis the limit as $b \rightarrow 0$ is taken, Model 9.1 is recovered.

9.3.2 Model 9.4 (Osaki and Vasudevan, 1972; Srinivasan and Sampath, 1976)

This is the continuous state analog of Model 8.5. It has the following features:

1) The e-events and the i-events form renewal processes with pdf $f(\cdot)$ and $g(\cdot)$ respectively.

2) An e-event increases the membrane potential $X(t)$ by a random amount, successive e-events resulting in such EPSPs that are iid with pdf $q(\cdot)$.

3) An i-event resets $X(t)$ to the rest level zero.

4) When $X(t)$ crosses K, the neuron fires and $X(t)$ is reset to zero.

In the absence of the i-events, the process $X(t)$ is simply the c-process of Section 9.1. However the arrival of an i-event complicates the process. Thus $X(t)$ can reach K by the cumulative effect of the e-events provided no i-event intervenes in the course of accumulation. It is obvious that the i-events make the output firing sequence a non-renewal process except when they are Poisson. Hence the interval pdf of the firing sequence does not characterise the process completely. Higher-order characteristics are therefore derived later in this section.

To derive an expression for the interval pdf, the following subsidiary functions are introduced.

$$\pi(t) = \lim_{\Delta,\Delta'\to 0} \text{Prob[the first i-event after the origin occurs in } (t,t+\Delta') \mid \text{a firing occurs in } (-\Delta,0)]/\Delta' . \qquad (9.3.20)$$

$$\pi_k(t) = \lim_{\Delta,\Delta',\Delta''\to 0} \text{Prob[the k-th i-event after the origin occurs in } (\tau,\tau+\Delta'), \text{ the first e-event after this occurs in } (t,t+\Delta'') \text{ and no firing has occurred in } (0,t) \mid \text{a firing occurs in } (-\Delta,0)]/\Delta'\Delta'' . \qquad (9.3.21)$$

$$p_k(t) = \lim_{\Delta,\Delta'\to 0} \text{Prob[a firing occurs in } (t,t+\Delta') \text{ and k i-events occur in } (0,t) \mid \text{a firing occurs in } (-\Delta,0)]/\Delta' . \qquad (9.3.22)$$

Then $\pi(\cdot)$, $\pi_k(\cdot,\cdot)$ and $p_k(\cdot)$ satisfy the following three lemmas respectively.

Lemma 9.3.3

$$\pi(t) = \gamma \int_0^\infty g(t+w)\, dw \Big[\int_K^\infty q(x)\, dx \Big\{ \int_W^\infty f(u)\, du$$

$$+ \sum_{m=1}^\infty \int_0^W F_m(u)\, du \int_{W-u}^\infty f(v)\, dv \Big\}$$

$$+ \sum_{n=1}^\infty \int_0^K q_n(x)\, dx \int_{K-x}^\infty q(y)\, dy \Big\{ \int_0^W f_n(u)\, du \int_{W-u}^\infty f(v)\, dv$$

$$+ \sum_{m=1}^\infty \int_0^W F_m(u)\, du \int_0^{W-u} f_n(z)\, dz \int_{W-u-z}^\infty f(v)\, dv \Big\} \Big] \qquad (9.3.23)$$

where γ is a normalising factor such that

$$\int_0^\infty \pi(t)\, dt = 1 . \qquad (9.3.24)$$

Lemma 9.3.4

$$\pi_1(t,\tau) = \pi(\tau)\ [f(t) + \sum_{m=1}^{\infty} f(t-u)\ f_m(u)\ du \int_0^K q_m(x)\ dx \tag{9.3.25}$$

and, for $k \geqslant 2$,

$$\pi_k(t,\tau) = \int_0^{\tau} \pi_{k-1}(t,u)\ g(\tau-u)\ du$$

$$+ \int_0^{\tau} f(t-u)\ du \int_0^{u} \pi_{k-1}(u,v)\ g(\tau-v)\ dv \int_0^K q(x)\ dx$$

$$+ \sum_{m=1}^{\infty} \int_0^{\tau} f(t-u)\ du \int_0^{u} f_m(u-v)\ dv$$

$$\int_0^{v} \pi_{k-1}(v,w)\ g(\tau-w)\ dw \int_0^K q_{m+1}(x)\ dx\ . \tag{9.3.26}$$

Lemma 9.3.5

$$p_0(t) = F(t) \int_t^{\infty} \pi(\tau)\ d\tau, \tag{9.3.27}$$

and for $k \geqslant 1$

$$p_k(t) = \int_0^{t} \pi_k(t,\tau)\ d\tau \int_{t-\tau}^{\infty} g(u)\ du \int_K^{\infty} q(x)\ dx$$

$$+ \sum_{m=1}^{\infty} \int_0^{t} f_m(t-v)\ dv \int_0^{v} \pi_k(v,\tau)\ d\tau \int_{t-\tau}^{\infty} g(u)\ du$$

$$\int_0^K q_m(x)\ dx \int_{K-x}^{\infty} q(y)\ dy, \tag{9.3.28}$$

where $F(\cdot)$ is the interval pdf in the c-process (see Theorem 9.2.1) i.e., equation (9.1.24), writing F for P.

Theorem 9.3.4 The pdf of the interval between two successive firings in Model 9.4 is given by

$$P(t) = \sum_{k=0}^{\infty} p_k(t). \qquad (9.3.29)$$

The above theorem can be simplified when either the e-events or the i-events form a Poisson process. The results are stated below in Theorems 9.3.5 and 9.3.6. The proofs are very similar to the proofs of Theorems 5.1.2 and 5.1.3 and are therefore omitted.

Theorem 9.3.5 In Model 9.4 when the i-events form a Poisson process with parameter μ, the Laplace transform of the pdf of the interval between two successive firings is given by

$$P^*(s) = \frac{F^*(s+\mu)\, f^*(s)\, [1 - f^*(s+\mu)]}{f^*(s+\mu)\,[1-f^*(s)] - [f^*(s) - f^*(s+\mu)]\, F^*(s+\mu)}. \qquad (9.3.30)$$

Corollary The mean length of the firing interval T in Model 9.4 is

$$\bar{T} = \frac{f^*(\mu)\,[1 - F^*(\mu)]}{\lambda_e[1-f^*(\mu)]\, F^*(\mu)}, \qquad (9.3.31)$$

where λ_e is the first order product density of the e-events.

Theorem 9.3.6 In Model 9.4 when the e-events form a Poisson process with parameter λ, the Laplace transform of the pdf of the interval between two successive firings is given by

$$P^*(s) = p_0^*(s) + \frac{p_{00}^*(s)\, B^*(s+\lambda)}{1 - C^*(s+\lambda)}, \qquad (9.3.32)$$

where $p_0(\cdot)$ is given by (9.3.27),

$$p_{00}(t) = F(t) \int_t^{\infty} g(u)\, du, \tag{9.3.33}$$

$$B(u) = \pi(u)\, A(u), \tag{9.3.34}$$

$$C(u) = g(u)\, A(u), \tag{9.3.35}$$

and

$$A(u) = 1 + \sum_{m=1}^{\infty} \frac{(\lambda u)^m}{m!} \int_0^K q_m(x)\, dx. \tag{9.3.36}$$

An alternative to Theorem 9.3.6 can be obtained by directly deriving an expression for $P(\cdot)$. To do this, first $\pi(\cdot)$ is obtained from (9.3.23) in simplified form as

$$\pi(t) = \frac{\gamma}{\lambda} \int_0^{\infty} g(t + w)\, h_c(w)\, dw \tag{9.3.37}$$

where $h_c(\cdot)$ is the renewal density of the firings in the c-process (Section 9.2.1). Next some additional information is obtained through the following lemmas.

<u>Lemma 9.3.6</u> When the e-events are Poisson with parameter λ, the pdf of the interval from a firing to the first i-event after this firing with no firing in between is given by

$$\pi_a(t) = \pi(t)\, [e^{-\lambda t} + \sum_{n=1}^{\infty} \frac{(\lambda t)^n e^{-\lambda t}}{n!} \int_0^K q_n(x)\, dx]. \tag{9.3.38}$$

<u>Lemma 9.3.7</u> When the e-events are Poisson with parameter λ, the pdf of the interval between two successive i-events with no firing in the interval is given by

$$\pi_b(t) = g(t)\, [e^{-\lambda t} + \sum_{n=1}^{\infty} \frac{(\lambda t)^n e^{-\lambda t}}{n!} \int_0^K q_n(x)\, dx]. \tag{9.3.39}$$

<u>Lemma 9.3.8</u> The renewal density corresponding to $\pi_b(\cdot)$ is given by

$$h_a(t) = \pi_b(t) + \int_0^t h_a(t)\, \pi_b(t-u)\, du. \qquad (9.3.40)$$

The interval pdf of the firing sequence is now given by

<u>Theorem 9.3.7</u> In Model 9.4, when the e-events are Poisson with parameter λ, the pdf of the interval between two successive firings is given by

$$P(t) = F(t) \int_t^\infty \pi(u)\, du + \int_0^t \pi_a(u)\, F(t-u)\, du \int_{t-u}^\infty g(v)\, dv$$
$$+ \int_0^t \pi_a(u)\, du \int_0^{t-u} h_a(v)\, F(t-u-v)\, dv \int_{t-u-v}^\infty g(w)\, dw, \qquad (9.3.41)$$

where $F(\cdot)$ is obtained from (9.1.24) writing F for P and putting $f(t) = \lambda e^{-\lambda t}$.

When both the e-events and the i-events are Poisson with parameter λ and μ respectively, $P^*(s)$ is given by (9.3.30) with

$$f^*(s) = \lambda/(s + \lambda), \qquad (9.3.42)$$

and the mean and the variance of the firing interval T are given by

$$\bar{T} = \frac{1 - F^*(\mu)}{\mu F^*(\mu)}, \quad \mu \neq 0 \qquad (9.3.43)$$
$$= -\dot{F}^*(\mu), \quad \mu = 0$$

and

$$\mathrm{Var}(T) = \frac{1 - F^{*2}(\mu) + 2\mu \dot{F}^*(\mu)}{(\mu F^*(\mu))^2}, \quad \mu \neq 0$$

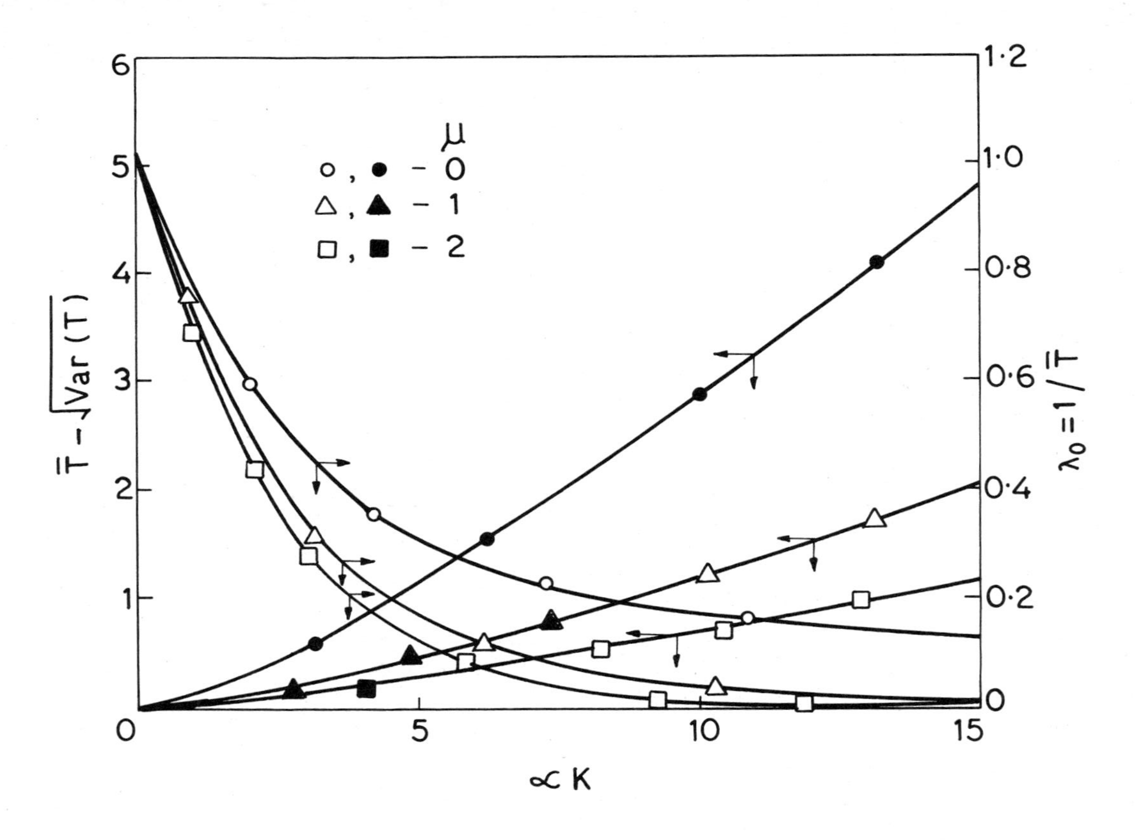

FIG. 9.3.1. MODEL 9.4 – GRAPHS OF $1/\bar{T}$ AND $\bar{T} - \sqrt{Var(T)}$ vs. αK

$$= \ddot{F}^*(0) - \dot{F}^{*2}(0), \quad \mu = 0. \tag{9.3.44}$$

Graphs relating $1/\bar{T}$ and $\bar{T} - \sqrt{Var(T)}$ to αK are presented in Figure 9.3.1. The significance of the latter set of graphs is explained in Section 10.3.

Higher order properties of the firing sequence (Srinivasan, 1976).

When the i-events do not form a Poisson process, higher order properties are required to characterise the sequence of firings. In the general case when the two sequences are renewal processes with pdf $f(\cdot)$ and $g(\cdot)$, the second order product density of the firings has been obtained by Srinivasan (1976). However an expression for the n-th order density can be found. This general result is stated below.

Consider the sequence of firings. Let $N(t)$ be the r.v. of the number of firings in $(0,t)$ and $I_B(\cdot)$ the backward recurrence time of the i-events. Let $h_c(\cdot)$ be the renewal density of the firings in the c-process.

Define

$$R(t,y) = \lim_{\Delta, \Delta' \to 0} \frac{\text{Prob}\,[N(t+\Delta) - N(t)=1,\ y < I_B(t) < y+\Delta']}{\Delta \Delta'} . \tag{9.3.45}$$

Then the following lemma can be written

Lemma 9.3.9

$$R(t,y) = \mu_i \lambda_e \int_y^\infty g(v)\, dv \int_0^y h_c(t-x)\, dx \int_x^\infty f(u)\, du, \tag{9.3.46}$$

where λ_e and μ_i are the stationary first order product densities of the e-and the i-events respectively.

Define

$$R^c(t,y \mid z) = \lim_{\Delta,\Delta',\Delta'' \to 0} \text{Prob}[N(t+\Delta) - N(t)=1,\ y < I_B(t) < y+\Delta' \mid N(0) - N(-\Delta'')=1,\ I_B(0) = z] / \Delta\Delta' \tag{9.3.47}$$

Lemma 9.3.10

$$R^c(t,y\,|\,z) = \frac{\int_0^t f(u)\, h_c(t-u)\, du \int_{t+z}^{\infty} g(v)\, dv}{\int_z^{\infty} g(v)\, dv} \cdot \delta(y-t-z)$$

$$+ \frac{h_i(t-y+z)}{\int_z^{\infty} g(v) dv} \int_0^u h_e(v)\, dv \int_{t-y}^{t} f(w-v)$$

$$h_c(t-w)\, dw \int_y^{\infty} g(w')\, dw', \qquad (9.3.48)$$

where $h_e(\cdot)$ and $h_i(\cdot)$ are the renewal densities of the e-events and the i-events respectively.

Using this the n-th order product density of the firings is obtained.

Theorem 9.3.8 The n-th order product density of the firings in Model 9.4 is given by

$$h_n(t_1, t_2 \cdots, t_{n-1}) = \int\int \cdots \int R(y)$$

$$R^c(t_1, y_1\,|\,y)\, R^c(t_2 - t_1, y_2\,|\,y_1) \cdots$$

$$R^c(t_{n-1} - t_{n-2}, y_{n-1}\,|\,y_{n-2})\, dy_1\, dy_2 \cdots dy_{n-1}. \qquad (9.3.49)$$

These results are much more general than those given by Srinivasan (1976).

9.4 Models with dependent interaction of e- and i-events

In Section 5.2, pre-inhibition models and in Section 8.4 discrete state models with dependent interaction of e-events and i-events have been considered. Similar models in continuous state can be constructed. Thus an e-event can trigger a sequence of i-events or an i-event can trigger a sequence of e-events. Since the analysis of these models is similar, only one model will be discussed in this section as an illustration.

9.4.1 Model 9.5

The model has the following properties:

1) The e-events form a renewal process with pdf $f(\cdot)$.
2) An e-event causes the membrane potential $X(t)$ to increase by a random amount, successive e-events resulting in such EPSPs that are iid with pdf $q(\cdot)$.
3) Every e-event triggers a sequence of i-events and stops the i-sequence started by the previous e-event. The i-events form a renewal process with pdf $g(\cdot)$, the interval from the e-event generating the sequence to the first e-event in this sequence also having a pdf $g(\cdot)$.
4) An i-event sets $X(t)$ to the rest level (zero).
5) When $X(t)$ crosses the threshold level K, the neuron fires and $X(t)$ is reset to zero.

Since the i-events are controlled by the e-events, the firing sequence is a renewal process. Using the distribution function defined by (9.2.3), one can write

$$h(x,t) = f(t)\, q(x) + \int_0^t f(t-u)\, du \int_{t-u}^{\infty} g(v)\, dv \int_0^x h(y,u)\, q(x-y)\, dy + q(x) \int_0^t f(t-u)\, du \int_0^{t-u} g(v)\, dv \int_0^K h(y,u)\, dy. \quad (9.4.1)$$

The pdf of the interval between two successive firings is now easily obtained.

Theorem 9.4.1 The pdf of the interval between two successive firings in Model 9.5 is given by

$$P(t) = f(t) \int_K^{\infty} q(x)\, dx$$

$$+ \int_0^t f(t-u)\, du \int_{t-u}^{\infty} g(v)\, dv \int_0^K h(x,u)\, dx \int_{K-x}^{\infty} q(y)\, dy$$

$$+ \int_K^{\infty} q(y)\, dy \int_0^t f(t-u)\, du \int_0^{t-u} g(v)\, dv \int_0^K h(x,u)\, dx. \tag{9.4.2}$$

The method of obtaining explicit formulas for $P(\cdot)$ is similar to that used in Section 9.2 and hence will not be pursued further.

9.5 Discussion

The continuous state models are quite comprehensive as many features of real neurons are represented in them. The distribution functions used are fairly general and there is, at least in theory, much flexibility so that a wide range of firing interval distributions can be generated. Many kinds of interactions have been modelled. Unfortunately except with very simple distributions (low-order Erlangian) these models present computational difficulties. Thus the method of solution of the integral equations formulated to describe the evolution with time of the membrane potential is based on the assumption that the pdfs can be approximated by gamma functions. It requires a preliminary solution of a polynomial with coefficients that are functions of the Laplace transform variable s. The restriction on analytical solution of polynomials immediately applies. Thus the highest degree of polynomial can be 4. This in turn restricts the order of the gamma functions assumed to approximate the pdfs. Polynomials of degree greater than 4 can be solved only numerically, and, since the coefficients involve the complex variable s, computation is a formidable task.

CHAPTER 10

REAL NEURONS AND MATHEMATICAL MODELS

In these notes, the spontaneous activity of several kinds of neurons has been modelled as a stochastic point process that results from the interaction of inputs to a neuron that are also stochastic processes. First related theory on which the models are based is presented and these are then enlarged in stages. Thus models with only one kind of input - excitatory - (Models 8.1, 8.2, 9.1 and 9.2) are frequently studied first before constructing more general models with two kinds of input - excitatory and inhibitory - (Models 8.3 to 8.6 and 9.3 to 9.5). Since the neuron receives inputs at its several synapses, spatial summation occurs at the membrane. This effect is modelled by superposition of input sequences (Models 4.1 to 4.3). The introduction of inhibition leads to a more ambitious modelling approach. Thus different kinds of interaction of excitatory and inhibitory inputs have been considered. This interaction may be (a) pre-synaptic (Models 5.1 to 5.4 and 9.2) or post-synaptic (Models 6.1, 6.2, 8.3 to 8.6 and 9.3 to 9.5) (b) independent (Models 5.1, 5.2, 6.1, 6.2, 8.3 to 8.5, 9.3 and 9.4) or dependent (Models 5.3, 5.4, 8.6 and 9.5). The inhibitory effect may be suppression of excitation (Models 5.1 to 5.4 and 9.2), repolarisation or hyperpolarisation of the membrane potential (Models 6.1, 6.2, 8.3 and 9.3) or reset of the potential to the rest level (Models 8.5, 8.6, 9.4 and 9.5). Throughout it has been assumed that the two input sequences are renewal processes. However, the firings need not constitute a renewal process if the inhibitory sequence is not a Poisson process (Models 8.5, 8.6 and 9.4). Properties of real neurons that have been taken into account are (a) integrate-and-fire behaviour (Models 6.1, 6.2, 7.6, 8.1 to 8.6 and 9.1 to 9.5), (b) variability of post-synaptic changes (Models 9.1 to 9.5), (c) membrane decay characteristic (Models 6.1, 6.2, 9.2 and 9.3) and (d) return of threshold to the normal level after a firing (Model 6.2). In the following sections a general discussion of the properties of real neurons in relation to mathematical models is provided.

10.1 Decay of the membrane potential

The membrane of the soma is similar to that of the axon. In the absence of stimuli, the membrane potential tends to return to the rest level. For this reason, the neuron is referred to as a leaky integrator — the post-synaptic potential changes when impulses arrive; in their absence, the membrane tends to return to equilibrium. It is important to note here that this decay is different from the decaying phase of the action potential waveform, the latter being described by the Hodgkin-Huxley equations. The former is a subthreshold characteristic of the membrane and it is not clear why this should be obtained from an approximation (Knight, 1972b) of the Hodgkin-Huxley equations which describe the membrane potential behaviour following stimulation to threshold level. Subliminal excitation is discussed by Ochs (1965, p. 45) and the decay following such excitation is roughly exponential.

Many attempts have been made to include some form of decay in neuron models. Thus, in Model 8.1, the EPSP lasts for a constant length of time. In Model 8.2, no two successive impulses may be separated by a time greater than a constant γ for a firing to occur through a bunching of N impulses. This constant γ is interpreted by Hochman and Fienberg (1971) as a kind of 'decay'. Barlow (1963) has raised objections to attributing a constant lifetime to impulses and suggests that the lifetime be treated as a random variable instead. Models 8.1 and 8.3 are partly based on this supposition. In Model 8.3, the 'decay' is introduced into the state equations (8.3.2 and 8.3.3) and the authors (Goel et al., 1972) argue that the average value of the potential in the absence of inputs (i.e., λ and μ are equal to zero in equations (8.3.11) and (8.3.12)) is an exponential function of time. In effect, what is done is addition of constants to the Poisson parameters to represent 'decay'.

The idea of a 'lifetime' of excitatory and inhibitory effects appears to be conceptually unsound. After an e-event (i-event) induces an EPSP (IPSP), the post-synaptic effect is summed by the cell so that one can talk only of the evolution of the membrane potential, since the EPSP or the IPSP no longer exists as an entity by itself. The decay is in the membrane potential and is due to the properties of the membrane. It is a deterministic rather than a stochastic phenomenon and has an exponential variation as mentioned above. Hence all the above attempts are only artifices.

The inclusion of exponential decay in diffusion models (Chapter 6) is, on the other hand, very realistic. In discrete state models, inclusion of any deterministic decay results in a model that is no longer discrete and gives rise to complications. In continuous state models, deterministic decay can be incorporated but only with a rather severe approximation, namely the linear, in order to give a solvable set of equations. This approximation, admittedly crude, has been made in Models 9.2 and 9.3. In counter models, decay can be brought in through the transfer function $g(\cdot)$ of the accumulator following the counter, but the resulting equations are non-linear and can be solved only numerically.

10.2 Hyperpolarisation of the membrane

It has been mentioned before that due to inhibition, the membrane potential can go below the rest level. Hyperpolarisation of the membrane potential does not appear to have been widely studied. While with depolarisation there is an upper limit which is the threshold of firing, what is the limit with hyperpolarisation? An obvious one is the breakdown potential of the membrane. However, in certain neurons, spikes due to inactivation occur (Grundfest, 1971). It remains to be seen what role such behaviour has in stochastic models.

Given the distribution of the inter-arrival times of the i-events and that of the IPSP, it is easy to prove, using Chebyshev's inequality (see, for example, Gnedenko (1969, p.202)), that the probability of the membrane potential reaching large negative values is negligibly small. In the light of the above, the inclusion by Gerstein and Mandelbrot (1964) of a lower bound below the rest level in their random walk model seems to be justified, despite Griffith's comments (Griffith, 1971, p.61). Mathematically this is necessary to avoid the pathological result of the total probability of reaching the threshold being less than 1. This latter difficulty also disappears when a decay parameter is included in the model ; see Section 6.3.

10.3 Refractoriness and threshold

Refractoriness in a real neuron has been described in Section 1.1.1. Absolute refractoriness can be taken into account in renewal models without much difficulty and is mathematically unimportant, though Griffith (1971, p.16) is of a contrary opinion. In a Poisson process the origin can be arbitrarily fixed; thus diffusion models (Chapter 6) easily take care of refractoriness. In a renewal process the dead time due to absolute refractoriness is easy to include when writing an integral relation. Thus, the first e-event after the absolute refractory period can be related to the last e-event occurring within the period through the renewal density. Considering the c-process in Section 9.2.1, for example, if an absolute refractory period T follows a firing, equation (9.1.24) can be written

$$P(t) = \int_0^T h_e(u)\, f(t-u)\, du \int_K^\infty q(x)\, dx$$

$$+ \sum_{m=1}^{\infty} \int_0^T h_e(u)\, du \int_T^t f(v-u)\, f_m(t-v)\, dv$$

$$\int_0^K q_m(x)\, dx \int_{K-x}^\infty q(y)\, dy, \qquad (10.3.1)$$

where $h_e(\cdot)$ is the renewal density of the e-events. It is the relative refractory period that is difficult to incorporate in stochastic models. This is because, during the period of relative refractoriness, the threshold is believed to return to the normal level from an infinite value. Simple variations like the Hagiwara threshold have been considered in some models (see, for example, Section 6.2.2). Generally, these result in equations that are very difficult to solve, and do not give closed form solutions. Others (for example, Holden (1976)) have considered the threshold to be a random variable, though it is not clear how far this is justified. However, a probabilistic interpretation of relative refractoriness can be given without disturbing the deterministic nature of the threshold returning to the normal level. In Models 9.2 and 9.4 graphs relating $\bar{T} - \sqrt{Var(T)}$ (the mean - the deviation from the mean of the interval between two successive firings) to the threshold level K have been given (Figures 9.2.1 and 9.3.1). Now $\sqrt{Var(T)}$ represents a deviation from the mean so that $\bar{T} - \sqrt{Var(T)}$ is a measure of the probability of occurrence of short intervals between two successive firings. A lower $\bar{T} - \sqrt{Var(T)}$ means a smaller probability of short intervals and vice versa. In Models 9.2 and 9.4, $\bar{T} - \sqrt{Var(T)}$ is always positive and increases as the threshold increases, i.e., with a larger threshold the probability of shorter intervals is smaller. Hence $\bar{T} - \sqrt{Var(T)}$ is a measure of 'probabilistic refractoriness'. Looking at the variation of the threshold, the neuron fires only rarely when the threshold is very high. But as the cell returns to the normal state after a firing, the threshold decreases and the probability of the membrane potential reaching the threshold increases. Now, because of absolute refractoriness intervals of zero length are not possible; an interval pdf therefore starts from a zero value. The initial variation of the pdf is similar to that of the rate of firing as the threshold returns to normal. Thus the variation with time of the threshold after absolute refractoriness and the relative refractoriness are reflected in the interval distribution.

10.4 Spatial summation

In most models, spatial summation of PSPs due to impulses arriving at the many synapses over the surface of the membrane is ignored or, when the number of synapses is large, approximated to a summation of PSPs due to a single sequence of Poisson events. However, it is possible that more than one stimulus may arrive at the neuron at the same time. Thus Eccles (1973, p. 71) discusses the linear summation of three different inputs at a motoneurone. In point process models it is assumed that multiple points (Srinivasan, 1969) do not occur. Such an assumption may be relaxed to take into account spatial summation. This way, the spatial variable need not be included in the model. There exist, on the other hand, neurons with single synapses (Bornstein, 1974) in which case simple stochastic models can be used to describe the firing process.

10.5 Other properties of neurons

It must be borne in mind that models have been constructed

on many simplifying assumptions. Thus properties of real neurons like (1) non-zero rise (fall) time of the membrane potential, i.e., the post-synaptic potential does not change instantaneously with the arrival of an excitatory (inhibitory) impulse, and (2) action potentials on dendrites are ignored. This is done to avoid complexities that cannot be treated mathematically. These properties have been taken into account in digital and analog simulation studies of neurons, and numerical estimates of parameters of neuron behaviour are available; see, for example, Perkel (1965) and Lewis (1968).

An interesting property of synapses that is yet to be considered in stochastic models is synaptic delay (Eccles, 1973, p. 47) The time of onset of an endplate potential after a nerve is stimulated is a random variable. Recordings made in the frog neuromuscular synapse show that the delay is made up of a constant value plus a random component which has a unimodal frequency distribution (Figure 10.5.1).

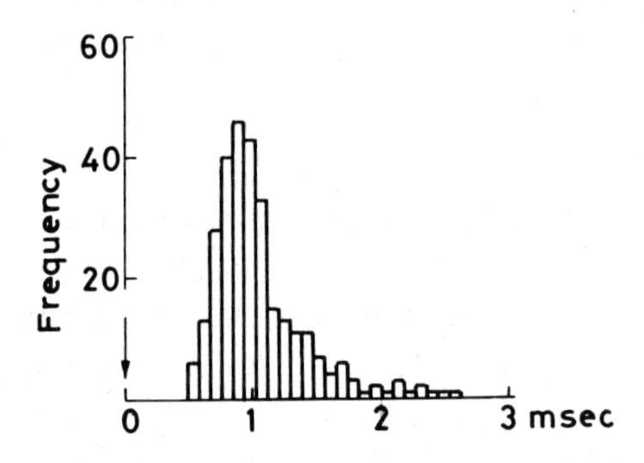

FIG.10·5·1 SYNAPTIC DELAY (After Eccles, 1973, p 47)

Evidently the time taken by an impulse to release transmitter molecules varies randomly. This is especially interesting in view of the growing belief that synapses can act as memory stores (see, for example, Fillenz (1972)).

The random nature of the synaptic delay can be mathematically considered random shifts of events in a point process. Ten Hoopen and Reuver (1967b) have considered a restricted case in which the spread of the delay distribution and of the interval pdf of the incident stream are much smaller than the mean interval of the incident sequence. A more general case without these restrictions has been treated by Srinivasan and Kumaraswamy (1971) using product densities, and an explicit expression for the characteristic functional of the delayed sequence obtained. However, it may be assumed that the random delay at a synapse does not alter the original sequence of stimuli, though the delayed sequence is no longer a simple renewal process but is a Markov renewal process. An interesting extension of such a scheme would be superposition of many such streams arriving at the many synapses and **study** of the threshold behaviour of the membrane potential, which is continually modified by the PSPs resulting from the superposed stream of delayed sequences.

10.6 The neuron as a black box

Generally, neuron models are aimed at generating output spike trains with characteristics that are similar to those of spike trains recorded in real neurons, by assuming certain properties of the input trains: a mechanism of interaction of these trains, the effect they have on the membrane potential, etc. In a different approach to the modelling problem, the output characteristics may be made part of the modelling scheme by postulating them as properties of the model and one of the other properties may be derived from the model. For example, Ricciardi and Capocelli (1972) have studied the properties of the input sequence that, when fed to a hypothetical neuron, generates an output spike train with known characteristics. One could attempt to do this for other properties, e.g., the decay rate of the membrane potential, the distribution of size of EPSPs and IPSPs, the number of synapses etc. Thus

1) one might wish to find out the frequency distributions of the size of the PSP that lead to output spike trains with known characteristics and compare them with those

obtained experimentally (see p.11).

2) the assumption that the input sequence is a Poisson process is based on the result of superposition of a large number of point processes. The parameter of the Poisson process depends on this number, which is also the number of synapses. Thus the number of synapses in a neuron can be estimated from a model with all the other characteristics known. Thus, equation (4.3.2) gives an estimate of this.

3) if a neuron has only one synapse, the threshold behaviour leads to an output firing rate that is very small. With a chain of such neurons, the information carried would be reduced to a negligible amount. If this does not happen, one may presume that (a) the input frequency is very high, (b) the mean of the PSP distribution is very large, or (c) the threshold is low. These characteristics can be predicted by the model.

10.7 Spike trains and renewal processes

The assumption that the input spike trains in a hypothetical neuron are renewal processes is basic to almost all stochastic models of neuron firing. This is not always justified, and many authors, for instance Stein (1972), refute the assumption. In view of this, it is necessary to develop models which relax it. This unfortunately results in excessive complexity. Simple models using other methods have been recently used to describe neural activity. These include the use of transfer functions (in the study of the Limulus eye by Dodge, Shapley and Knight (1970)) and coherence functions (in the study of cat muscle receptors by Stein (1972)). These are by and large empirical in their formulation and are models based on the results of a study of stimulated activity; and rest their faith on the assumption that biological systems are linear systems over a fairly wide range of behaviour.

There have been no serious attempts to evolve models using general point process inputs. In this the product density approach may be useful, because it can be used in studying non-Markovian and non-stationary processes as well. It must be admitted that it is not going to be easy.

10.8 Conclusion

Many types of models of spike trains generated by spontaneous activity in single neurons have been described in these notes. It has been assumed that the neuron is isolated; the influence of neighbouring neurons is brought in indirectly in some of the models (Models 5.3, 5.4, 8.6 and 9.5). But then, the nervous system is a large network of neurons and the interaction between neurons is quite complex. While deterministic behaviour of such nets has been studied by many authors, notably Caianiello (1961), whose paper is really an attempt at a global theory of thought processes, and Reichardt and MacGinitie (1962), who have modelled visual processes, the spontaneous activity of neural nets has only recently been studied by Knight (1972b) and Taylor (1972), though computer simulations are available (Moore, 1971). This is undoubtedly due to the complexity of biological systems. Nevertheless, such studies would be interesting because of their likely propensity to describe spontaneous behaviour in large nets like hallucination and dreaming and pathological conditions like epileptic attacks and catatonic seizures.

REFERENCES

Abramowitz, M. and Stegun, I.A. (1965). Handbook of mathematical functions. Washington, D.C.: National Bureau of Standards.

Barlow, H.B. (1963). " The information capacity of nervous transmission". Kybernetik 2, 1.

Bishop, P.O., W.R.Levick and W.O.Williams (1964). "Statistical analysis of the dark discharge of lateral geniculate neurons". J.Physiol. 170, 598-612.

Bornstein, J.C. (1974). "Multiquantal release of ACh in mammalian ganglia". Nature 248, 529-31.

Bush, B.M.H. and A.Roberts (1968). "Resistance reflexes from a crab muscle receptor without impulses". Nature 218, 1171-3.

Caianiello, E.R. (1961). "Outline of a theory of thought processes and thinking machines". J.Theor.Biol. 2, 204-35.

Capocelli, R.M. and L.M.Ricciardi (1971). "Diffusion approximation and first-passage time problem for a model neuron". Kybernetik 8, 214-23.

del Castillo, J. and B.Katz (1956). Biochemical aspects of neuromuscular transmission. Progr.Biophys.Biochem. 6, 121-70.

Clay, J.R. and N.S.Goel (1973). "Diffusion models for firing of a neuron with varying threshold". J.Theor.Biol. 39, 633-44.

Coleman, R. and J.L.Gastwirth (1969). "Some models for interaction of renewal processes related to neuron firing". J.Appl.Prob. 6, 38-58.

Copson, E.T. (1935). An introduction to the theory of functions of a complex variable. London: Oxford University Press.

Cox, D.R. (1962). Renewal theory. London: Methuen.

Cox, D.R. and H.D.Miller (1965). Theory of stochastic processes. New York: Wiley.

Dodge, Jr.,F.A., B.W.Knight and J.Toyoda (1968). "Voltage noise in Limulus visual cells". Science 160, 88-90.

Dodge, F.A., R.M.Shapley and B.W.Knight (1970). "Linear systems analysis of the Limulus retina". Behav.Sci. 15, 24-36.

Eccles, J.C. (1964). The physiology of synapses. Heidelberg: Springer.

Eccles, J.C. (1973). The understanding of the brain. New York: McGraw-Hill.

Ekholm, A. (1972). 'A generalisation of the two-state two-interval semi-Markov model'. In - P.A.W.Lewis (ed.). Stochastic point processes. New York: Wiley Interscience.

Eyzaguirre, C. (1969). Physiology of the nervous system. Chicago: Year Book Medical Publishers.

Feller, W. (1960). An introduction to probability theory and its applications. Vol.1. Bombay: Asia Publishing House.

Fetz. E.E. and G.L.Gerstein (1963). "An RC model for spontaneous activity of single neurons". Quart.Prog.Rep. 71, Res.Lab. Electronics MIT, 249-57.

Fienberg, S.E. (1974). "Stochastic model for single neuron firing trains: a survey". Biometrics 30, 399-427.

Fienberg, S.E. and Hochman, H.G. (1972). "Modal analysis of renewal models for single neuron discharge". Kybernetik 11, 292-7.

Fillenz, M. (1972) " Hypothesis for a neuronal mechanism involved in memory". Nature 238, 41-3.

Furshpan, E.J. (1964). " 'Electrical transmission' at an excitatory synapse in a vertebrate brain". Science 144, 878-80.

Gerstein, G.L. and B.Mandelbrot (1964). " Random walk models for the spike activity of a single neuron" . Biophys.J. 4, 41-67.

Gluss, B. (1967). "A model for neuron firing with exponential decay of potential resulting in diffusion equations for probability density". Bull. Math. Biophys. 29, 233-43.

Gnedenko, B.V. (1969). The theory of probability (tr. from Russian by G.Yankovsky). Moscow: Mir Publishers.

Goel, N.S., N.Richter Dyn and J.R.Clay (1972). "Discrete stochastic models for firing of a neuron". J.Theor.Biol. 34, 155-84.

Griffith, J.S. (1971). Mathematical neurobiology. New York: Academic Press.

Grundfest, H. (1971). 'The varieties of excitable membranes'. In - Adelman, W.J. (ed.). Biophysics and physiology of excitable membranes. New York: Van Nostrand Reinhold.

Hin, P.A. (1974). 'Interaction of stationary stochastic point processes related to neuron firing'. M.Sc. dissertation, University of Malaya, Kuala Lumpur.

Hirst, G.D.S. and H.C.McKirdy (1974). " Presynaptic inhibition at mammalian peripheral synapse?". Nature 250, 430-1.

Hochman, H.G. and S.E.Fienberg (1971). "Some renewal process models for single neuron discharge". J.Appl.Prob. 8, 802-8.

Hodgkin, A.L. and A.F.Huxley (1952a). "Currents carried by sodium and potassium ions through the membrane of the giant axon of Loligo" . J.Physiol. 116, 449-72.

Hodgkin, A.L. and A.F.Huxley (1952b). " The components of membrane conductance in the giant axon of Loligo" . ibid., 473-96.

Hodgkin, A.L. and A.F.Huxley (1952c). " The dual effect of membrane potential on sodium conductance in the giant axon of Loligo" ibid., 497-506.

Hodgkin, A.L. and A.F.Huxley (1952d). " A quantitative description of membrane current and its application to conductance and excitation in nerve ". J.Physiol. 117, 500-44.

Holden, A.V. (1976). Models of the stochastic activity of neurones. Lecture Notes in Biomathematics 12. Berlin: Springer-Verlag.

Katz, B. (1966). Nerve, muscle and synapse. New York: McGraw-Hill.

Khintchine, A.Y. (1960). Mathematical models in the theory of queueing. London: Griffin.

Knight, B.W (1972a). 'Some point processes in motor and sensory neurophysiology'. In-Lewis, P.A.W.(ed.). Stochastic point processes. New York: Wiley Interscience.

Knight, B.W. (1972b). " Dynamics of encoding in a population of neurons " . J.gen.Physiol. 59, 734-66.

Kuffler, S.W., R.Fitzhugh and H.B.Barlow (1957). "Maintained activity in the cat's retina in light and darkness". J.gen. Physiol. 40, 683-702.

Langer, R.E. (1954). A first course in ordinary differential equations. New York: Wiley.

Lawrance, A.J. (1970). "Selective interaction of a stationary point process and a renewal process" . J.Appl.Prob. 7, 483-9.

Lee, P.A. (1974). "Output response of non-linear switching element with stochastic dead time". Kybernetik 15, 187-191.

Leslie, R.T. (1969). "Recurrence times of clusters of Poisson points" . J.Appl.Prob. 6, 372-88.

Levick, W.R. and W.O.Williams (1964). "Maintained activity of lateral geniculate neurons in darkness". J.Physiol. 170, 582-97.

Lewis, E.R. (1968). "Using electronic circuits to model simple neuroelectric interactions". Proc. IEEE 56, 931-49.

Lindley, D.V. (1969). Introduction to probability and statistics - Part I. London: Cambridge University Press.

Martin, A.R. and G.Pilar (1964). "Quantal components of the synaptic potential in the ciliary ganglion of the chick". J.Physiol. 175, 1-16.

Molnar, C.E. and R.R.Pfeiffer (1968). "Interpretation of spontaneous spike discharge patterns of neurons in the cochlear nucleus". Proc. IEEE 56, 993-1004.

Moore, G. (1971). 'Bioengineering techniques for the nervous system'. In - Brown, J.H.U., J.H.Jacobs and L.Stark (eds.). Biomedical engineering, 97-121. Philadelphia: Davis.

Ochs, S. (1965). Elements of neurophysiology. New York: Wiley.

Osaki, S. (1971). "Notes on renewal processes and neuronal spike trains". Math.Biosc. 12, 33-9.

Osaki, S. and R.Vasudevan (1972). "On a model of neuronal spike trains". Math.Biosc. 14, 337-41.

Penner, M.J. (1972). "Neural or energy summation in a Poisson counting model". J.Math.Psychol. 9, 286-93.

Perkel, D.H. (1965). 'Application of a digital computer simulation of a neural network'. In - M.Maxfield, A.Callahan and L.J. Fogel (eds.). Biophysics and cybernetics systems, 37-51. Washington, D.C.: Spartan.

Poggio, G.F. and L.J.Viernstein (1964). "Time series analysis of impulse sequences of thalamic sensory neurons". J.Neurophysiol. 27, 517-45.

Rade, L. (1972), "A model for interaction of a Poisson and a renewal process and its relation with queueing theory". J. Appl.Prob. 9, 451-6.

Ramakrishnan, A. and P.M.Mathews (1953). "On a stochastic problem relating to counters". Phil. Mag. 44, 1122-8.

Rapoport, A. (1950). "Contribution to the probabilistic theory of neural nets". Bull. Math. Biophys. 12, 187-97.

Ratliff, F. (1961). 'Inhibitory interaction and the detection and enhancement of contours'. In - W.A.Rosenblith (ed.). Sensory communication. Cambridge (Mass.): MIT Press.

Reichardt, W. and G.MacGinitie (1962). "Zur theorie der lateralen inhibition". Kybernetik 1, 155-65.

Ricciardi, L.M. and R.M.Capocelli (1972). "On the inverse of the first-passage time probability problem". J.Appl.Prob. 9, 270.

Roy, B.K. and D.R.Smith (1969). "Analysis of the exponential decay model of the neuron showing frequency threshold effects". Bull-Math.Biophys. 31, 341-57.

Rushton, W.A.H. (1961). 'Peripheral coding in the nervous system'. In - W.A.Rosenblith (ed.). Sensory communication, 168-81. Cambridge (Mass.): MIT Press.

Sabah, N.H. and J.T.Murphy (1971). "A superposition model of the spontaneous activity of cerebellar Purkinje cells". Biophys.J. 11, 414-28.

Skvaril, J., T.Radil-Weiss, Z. Bohdanecky and J.Syka (1971). "Spontaneous discharge patterns of mesencephalic neurons: interval histogram and mean interval relationship". Kybernetik 9, 11-15.

Sneddon, I.N. (1957). Elements of partial differential equations. New York: McGraw-Hill.

Srinivasan, S.K. (1969). Stochastic theory and cascade processes. New York: American Elsevier.

Srinivasan, S.K. (1974a). Stochastic point processes and their applications. London: Griffin.

Srinivasan, S.K. (1974b). "Analytical solution of a finite dam governed by a general input". J.Appl.Prob. 11, 134-44.

Srinivasan, S.K. (1976). "A stochastic model of neuronal firing". To be published in Math.Biosc.

Srinivasan, S.K. and Kumaraswamy (1971). "Delayed events and cluster processes". J.Math.Phys.Sci. 5, 229-38.

Srinivasan, S.K. and K.M.Mehata (1976). Stochastic processes. New Delhi: Tata McGraw-Hill

Srinivasan, S.K. and G.Rajamannar (1970a). "Selective interaction between two independent stationary recurrent point processes". J.Appl.Prob. 7, 476-82.

Srinivasan, S.K. and G.Rajamannar (1970b). "Renewal point processes and neuronal spike trains". Math.Biosc. 6, 331-5.

Srinivasan, S.K. and G.Rajamannar (1970c). "Counter models and dependent renewal point processes related to neuronal firing". ibid. 7, 27-39.

Srinivasan, S.K., G.Rajamannar and A.Rangan (1971). "Stochastic models for neuronal firing". Kybernetik 8, 188-93.

Srinivasan, S.K. and G.Sampath (1975). "A neuron model with pre-synaptic deletion and post-synaptic accumulation, decay and threshold behaviour". Biol.Cybernetics 19, 69-74.

Srinivasan, S.K. and G.Sampath (1976). "On a stochastic model for the firing sequence of a neuron". Math.Biosc. 30, 305-23.

Srinivasan, S.K. and R.Vasudevan (1969). "On the response output from nonlinear switching elements with different types of finite dead-times". Kybernetik 6, 121-4.

Stein, R.B. (1965). "A theoretical analysis of neuronal variability". Biophys. J. 5, 173-94.

Stein, R.B. (1967). "Some models of neuronal variability". ibid. 7, 37-67.

Stein, R.B. (1972). 'The stochastic properties of spike trains recorded from nerve cells'. In - P.A.W.Lewis (ed.). Stochastic point processes, 700-31. New York: Wiley-Interscience.

Tasaki, I. (1968). Nerve excitation. Springfield: Charles C. Thomas.

Taylor, J.G. (1972). "Spontaneous behaviour in neural networks". J.Theor.Biol. 36, 513-28.

Ten Hoopen, M. (1966a). "Multimodal interval distributions". Kybernetik 3, 17-24.

Ten Hoopen, M. (1966b). "Probabilistic firings of neurons considered as a first passage time problem". Biophys.J. 6, 435-51.

Ten Hoopen, M. (1967). "Pooling of impulse sequences, with emphasis on applications to neuronal spike data". Kybernetik 4, 1-10.

Ten Hoopen, M. and H.A.Reuver (1965a). "Remark on the input-output relation of formalised neurons in the case of randomised stimuli" Bull. Math. Biophys. 27, 145-52.

Ten Hoopen, M.and H.A.Reuver (1965b). "Selective interaction of two recurrent processes". J.Appl.Prob. 2, 286-92.

Ten Hoopen, M. and H.A.Reuver (1967a). "On a first passage problem in stochastic storage systems with total release". J.Appl.Prob. 4, 409-12.

Ten Hoopen, M.and H.A.Reuver (1967b). "Analysis of sequences of events with random displacements applied to biological systems". Math.Biosc. 1, 599-617.

Ten Hoopen, M. and H.A.Reuver (1968). "Recurrent point processes with dependent interference with reference to neuronal spike trains". Math.Biosc. 2, 1-10.

Additional references

Arrow, K.J., S.Karlin and H.Scarf (1958). Studies in the mathematical theory of inventory and production. Stanford: University Press.

Bartlett, M.S. (1966). An introduction to stochastic processes. Cambridge: University Press.

Basawa, I.V. (1971). "Some models based on the interaction of two independent Markovian point processes". J.Appl.Prob. 8, 193-7.

Bayley, E.J. (1968). "Spectral analysis of pulse frequency modulation in the nervous system". IEEE Trans. BME 15, 257-65.

Bellman, R.E. and T.E.Harris (1948). "On the theory of age dependent stochastic branching processes". Proc.Nat.Acad.Sci. (U.S.A.) 34, 601-4.

Bharucha-Reid, A.T. (1960). Elements of the theory of Markov processes and their applications". New York: McGraw-Hill.

Burns, B.D. (1968). The uncertain nervous system. London: Arnold.

Çinlar, E. (1972). 'Superposition of point processes'. In — P.A.W. Lewis (ed.). Stochastic point processes. New York: Wiley-Interscience.

Çinlar, E. (1975). Introduction to stochastic processes. Englewood-Cliffs: Prentice Hall.

Cooke, I. and M.Lipkin (1972) (eds.). Cellular neurophysiology - a source book. New York: Holt Rinehart and Winston.

Cox, D.R. and P.A.W.Lewis (1966). The statistical analysis of series of events. London: Methuen.

Hawkes, A.G. (1970). "Bunching in a semi-Markov process". J.Appl. Prob. 7, 175-82.

Kabe, D.G. (1967). "A note on some distributions for nonlinear switching elements with finite dead time". Kybernetik 3, 285-7.

Kendall, D.G. and E.F.Harding (1973) (eds.). Stochastic analysis. London: Wiley.

Kendall, D.G. and E.F.Harding (1973) (eds.). Stochastic geometry. London: Wiley.

Kingman, J.F.C. (1972). Regenerative phenomena. London: Wiley.

Lewis, P.A.W. (1972). Stochastic point processes — statistical analysis, theory and applications. New York: Wiley-Interscience.

Marmarelis, P.Z. and G.D.McCann (1973). "Development and application of white noise modelling techniques for studies of insect visual nervous system". Kybernetik 12, 74-89.

Moran, P.A.P. (1959). The theory of storage. London: Methuen.

Murthy, V.K. (1974). General point processes — applications to structural fatigue, bioscience and medical research. Reading: Addison-Wesley.

Parzen, E. (1959). Time series analysis papers. San Francisco: Holden-Day.

Perkel, D.H., G.L.Gerstein and G.P.Moore (1967). "Neuronal spike trains and stochastic point processes - Parts 1 and 2". Biophys. J. 7, 391-440.

Plonsey, R. (1969). Bioelectric phenomena. New York: McGraw-Hill.

Prabhu, N.U. (1965). Queues and inventories. New York: Wiley.

Ramakrishnan, A. (1958). 'Probability and stochastic processes'. In — Handbuch der Physik, Vol.3. Berlin: Springer-Verlag.

Ricciardi, L.M. and F.Esposito (1966). "On some distribution functions for nonlinear switching elements with finite dead time". Kybernetik 3, 148-50.

Saaty, T.L. (1961). Elements of queueing theory. New York: McGraw-Hill.

Seal, H.L. (1969). Stochastic theory of risk business. New York: Wiley.

Smith, W. (1973). "Shot noise generated by a semi-Markov Process". J.Appl.Prob. 10, 685-90.

Srinivasan, S.K. and R.Subramanian (1969). "Queueing theory and imbedded renewal processes". J.Math.Phys.Sci. 3, 221-44.

Ten Hoopen , M. and H.A.Reuver (1967). "Interaction between two independent recurrent time series". Inform.Control 10, 149-58.

Vere-Jones, D. (1966). "Simple stochastic models for the release of quanta of transmitter from a nerve terminal". Austral.J.Stat. 8, 53-63.

INDEX